精细化工专业新工科系列教材

精细化工专业英语

ENGLISH FOR FINE CHEMICALS

朱为宏　郭志前　主编

化学工业出版社
·北京·

内 容 简 介

《精细化工专业英语》是华东理工大学精细化工专业新工科系列教材之一。全书共分为5个部分，23个单元。第一部分着重介绍元素周期表、无机与有机化合物的命名方法和性质；第二部分主要讲述有机化合物性质表征方法，包括紫外-可见光谱、红外光谱、核磁共振、质谱等；第三部分为精细有机合成典型单元操作过程，涉及重结晶、蒸馏、分离、合成放大等；第四部分对精细化学品种类进行了介绍，包括香料、化妆品、染料、手性药物等；第五部分介绍精细化学品领域的最新进展和应用，包括荧光染料及荧光化学传感器、有机光致变色染料、染料敏化太阳能电池、分子机器/分子逻辑计算以及碳中和战略简介。

《精细化工专业英语》每单元配有单词和词组表，并作了必要的注释，便于学生自学。本书内容既注重对专业知识面的覆盖，又顾及专业的历史发展，并反映当前新工科的发展趋势和若干重要科技领域（如能源、环保等）的新需求，可作为高校精细化工、应用化学及相关专业的教材，也可供相关领域科技人员参考。

图书在版编目（CIP）数据

精细化工专业英语/朱为宏，郭志前主编. —北京：化学工业出版社，2022. 8
精细化工专业新工科系列教材
ISBN 978-7-122-41898-2

Ⅰ. ①精… Ⅱ. ①朱… ②郭… Ⅲ. ①精细化工—英语—高等学校–教材 Ⅳ. ①TQ39

中国版本图书馆 CIP 数据核字（2022）第 131060 号

责任编辑：任睿婷　吕　尤　徐雅妮　　装帧设计：李子姮
责任校对：宋　玮

出版发行：化学工业出版社（北京市东城区青年湖南街 13 号　邮政编码 100011）
印　　装：北京印刷集团有限责任公司
787mm × 1092mm　1/16　印张 11¾　字数 291 千字　　2023 年 2 月北京第 1 版第 1 次印刷

购书咨询：010-64518888　　售后服务：010-64518899
网　　址：http://www.cip.com.cn
凡购买本书，如有缺损质量问题，本社销售中心负责调换。

定　　价：39.00 元

前　言

精细化工专业是化学、化工、材料等多学科的交叉融合。随着现代科学技术的进步和发展，世界精细化工行业也发生了巨大的变化，展现出以分子工程为基础，强调在分子尺度上对化工产品性能进行精确定制研究和开发的新趋势。围绕国家重大战略开展“新工科”建设是主动应对新一轮科技革命与产业变革的战略行动，建立新工科课程体系势在必行。

精细化工专业英语既是传播专业知识的媒介，又是通向精细化工世界的桥梁。通过精细化工专业英语的学习，学生将进一步了解专业知识，熟悉专业文章的英文表达，提高化工英文文献的阅读能力。本书分为 5 个部分，以元素周期表认知、基本无机和有机化合物命名、有机化合物性质表征分析、精细有机合成典型单元操作、精细化学品介绍、精细化学品在新兴领域的应用进展等为主线展开，主要内容参考了原版英文教材、科技报告、专著及最新的专业期刊论文，期望通过英语文献学习加强读者对基本专业英语词汇的认知和理解。

本书由华东理工大学朱为宏、郭志前主编，张志鹏、吴永真、燕宸旭等参加了全书的编写工作。全书由朱为宏、郭志前统稿，燕宸旭负责练习和注释的编写。本教材在成书过程中得到了化学工业出版社和华东理工大学教务处的大力支持，在此表示衷心的感谢。

由于编者水平有限，疏漏和不妥之处在所难免，恳请读者批评指正。

编　者

2022 年 6 月

Contents

PART 1

Introduction of General Chemicals

UNIT 1 Elements and the Periodic Table

Chemistry is the science that deals with matter: the structure and properties of matter and the transformations from one form of matter to another. The world around us is made of chemicals. Our food, our clothing, the buildings in which we live are all made of chemicals. Our bodies are made of chemicals, too. To understand the human body, its diseases, and its cures, we must know all we can about those chemicals. A chemical element is a building block of matter.

Many of the elements found throughout nature are also found within the body. There are several ways to consider the composition of the human body, including elements, type of molecules, or type of cells. Water is the most abundant chemical compound in living human cells, accounting for 65% to 90% of each cell. Therefore, it is not surprising that most of a human body's mass is oxygen. It is also present between cells. For example, blood and cerebrospinal fluid are mostly water. Carbon, the basic unit for organic molecules, comes in second. Organic compounds include fat, proteins, carbohydrates, and nucleic acids. The percentage of fat varies from person to person, but even an obese person has more water than fat. In a lean male, the percentages of proteins and water are comparable. Muscles, including the heart, contain a lot of proteins with different functions. Hair and fingernails are proteins. Skin contains a large amount of proteins, too. Although humans use glucose as an energy source, there is not that much of it free in the bloodstream at any given time. Sugar and other carbohydrates only account for about 1% of body mass. Fig.1.1 is the chemical composition of the average adult human body in terms of elements and compounds. The mass of the human body is made up of just four elements: oxygen, carbon, hydrogen, and nitrogen.

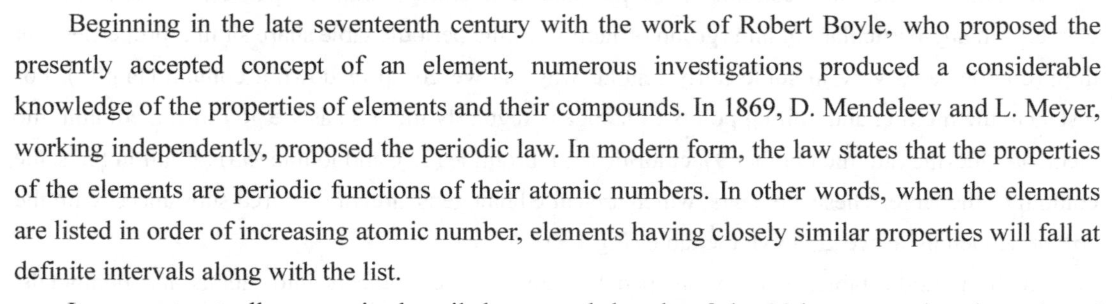

Beginning in the late seventeenth century with the work of Robert Boyle, who proposed the presently accepted concept of an element, numerous investigations produced a considerable knowledge of the properties of elements and their compounds. In 1869, D. Mendeleev and L. Meyer, working independently, proposed the periodic law. In modern form, the law states that the properties of the elements are periodic functions of their atomic numbers. In other words, when the elements are listed in order of increasing atomic number, elements having closely similar properties will fall at definite intervals along with the list.

It was not actually recognized until the second decade of the 20th century that the order of elements in the periodic system is that of their atomic numbers, the integers of which is equal to the positive electrical charges of the atomic nuclei expressed in electronic units. In subsequent years great

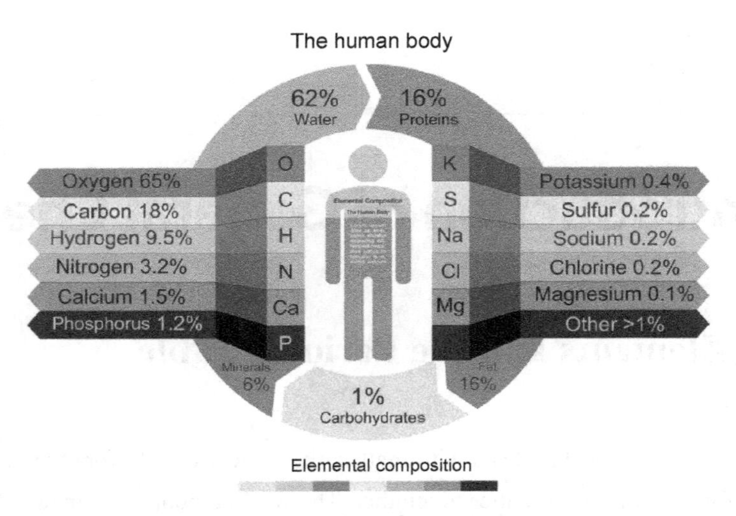

Fig.1.1 Illustration of the makeup of human body

progress was made in explaining the periodic law in terms of the electronic structure of atoms and molecules. This clarification has increased the value of the law, which is used as much today as it was at the beginning of the 20th century, when it expressed the only known relationship among the elements.

Element groups are the columns in the periodic table. They include alkali metals, alkaline earth metals, transition metals, basic metals, metalloids, halogens, and noble gases. The two rows of elements located below the main body of the periodic table are a special group of transition metal called the rare earth elements. The lanthanides are the elements in the top row of the rare earths. The actinides are elements in the bottom row.

Each horizontal row of elements constitutes a period. It should be noted that the lengths of periods vary. There is a very short period containing only 2 elements, followed by two short periods of 8 elements each, and then two long periods of 18 elements each. The next period includes 32 elements, and the last period is apparently incomplete. With this arrangement, elements in the same vertical column have similar characteristics. These columns constitute chemical families or groups. The groups headed by the members of the two 8-elements are designated as main group elements, and the members of the other groups are called transition or inner transition elements.

Elements in the same column of the periodic table show similar properties. Not only do the elements in any particular column (group or family) of the periodic table share similar properties, but the properties also vary in some fairly regular ways as we go up or down a column (family). For instance, the melting and boiling points of halogens regularly increase as we go down a column: the elements fluorine (atomic number 9), chlorine (17), bromine (35), and iodine (53) all fall in the same column of the table. These elements, which are called halogens, are all coloured substances, with the colour deepening as we go down the table (Fig.1.2).

In the periodic table, a heavy stepped line divides the elements into metals and nonmetals. Elements to the left of this line (except hydrogen) are metals, while those to the right are nonmetals. This division is for convenience only: elements bordering the line (metalloids) have the characteristic of both metals and nonmetals. It may be seen that most of the elements, including all transition and inner transition elements, are metals.

Except for hydrogen, a gas, the elements of group IA make up the alkali metal family. They are very reactive metals, and they are never found in the elemental state in nature. However, their compounds are widespread. All the members of the alkali metal family form ions having a charge of 1+ only. In contrast, the elements of group IB — copper, silver, and gold — are comparatively inert. They are similar to the alkali metals in that they exist as 1+ ions in many of their compounds. However, as is the characteristic of most transition elements, they form ions having other charges as well.

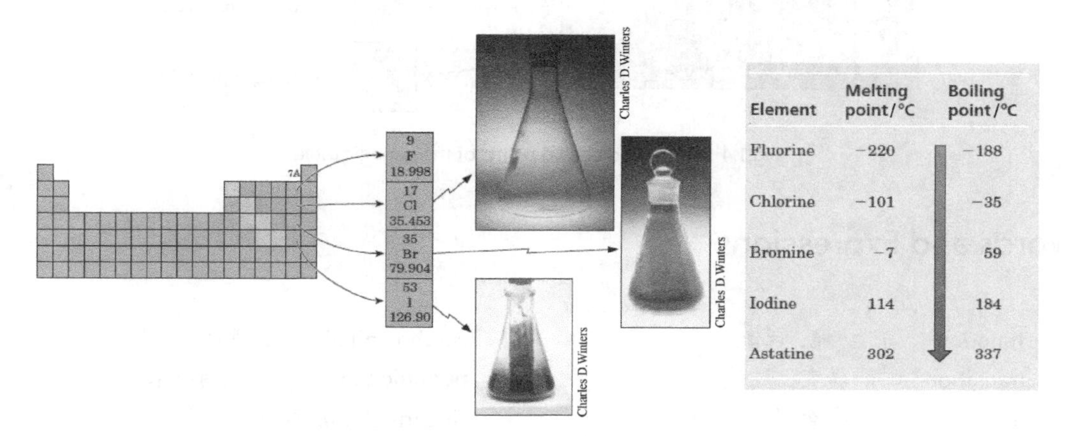

Element	Melting point/°C	Boiling point/°C
Fluorine	−220	−188
Chlorine	−101	−35
Bromine	−7	59
Iodine	114	184
Astatine	302	337

Fig.1.2 Melting and boiling points of halogens (Group VIIA elements)

Element symbols are the abbreviations of elements' name. In some cases, the abbreviations come from the elements' Latin name. Each symbol is either one or two letters in length. Usually, the symbol is an abbreviation of the element name, but some symbols refer to older names of the elements (for example, the symbol for silver is Ag, which refers to its old name, argentum). Each element on the periodic table is represented by the atomic symbol, the atomic number in the upper lefthand corner.

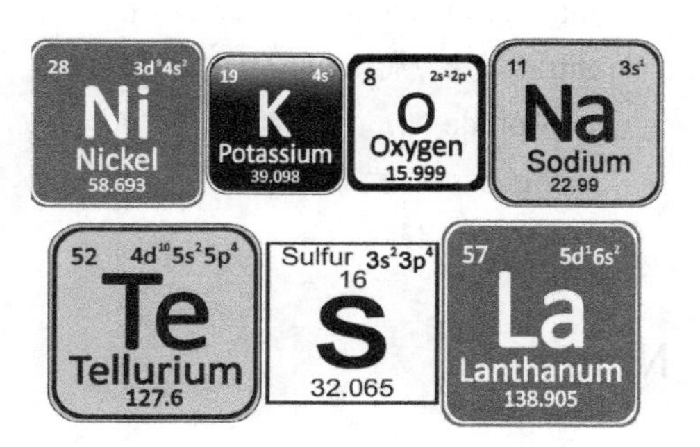

Fig.1.3 Structure of the periodic table showing electron shells

The periodic table was originally constructed on the basis of trends (periodicity) in physical and chemical properties. With an understanding of electron configurations, chemists realized that the periodicity in chemical properties could be understood in terms of the periodicity in ground-state electron configuration. The periodic table works because elements in the same column have the same ground-state electron configuration in their outer shells (Fig.1.3).

The periodic table is so useful that it hangs in nearly every chemistry classroom and chemical laboratory throughout the world. What makes it so useful is that it correlates a vast amount of data about elements and their compounds, and allows us to make many predictions about both chemical and physical properties. For example, if you are told that the boiling point of germane (GeH_4) is −88℃ and that of methane (CH_4) is −164℃, could you predict the boiling point of silane (SiH_4)? The position of silicon in the table, between germanium and carbon, might lead you to a prediction of about −125℃. The actual boiling point of silane is −112℃, not far from this prediction (Fig.1.4).

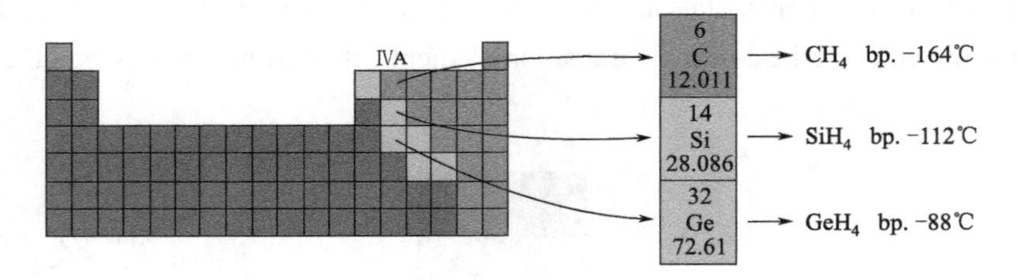

Fig.1.4 Prediction boiling points of the periodic table

Words and Expressions

function *n.* 功能；函数

transition *n.* 转变；过渡

metalloid *n.* 半金属

alkaline earth metal 碱土金属

main group element 主族元素

nitride *n.* 氮化物

phosphide *n.* 磷化物

halogen *n.* 卤素

silver *n.* 银

cerebrospinal *adj.* 脑脊髓的

periodic table [元素]周期表

fingernail *n.* 指甲

carbohydrate *n.* 糖类

rare earth element 稀土元素

silane *n.* 硅烷

tellurium *n.* 碲

lanthanum *n.* 镧

Notes

➢ Beginning in the late seventeenth century with the work of Robert Boyle, who proposed the presently accepted concept of an element, numerous investigations produced a considerable knowledge of the properties of elements and their compounds.

参考译文：早在17世纪末期，罗伯特·波义尔就开始了这项工作，他提出了现在公认的元素概念，通过大量的研究使我们对元素及其化合物的性质有了充足的了解。

➢ They form such a wide variety of compounds that it is not practical at this point to present any examples as being typical of the respective groups.

参考译文：它们形成了如此众多的化合物种类，以致无法举出任何能表现各组元素典型变化的例子。

Exercises

1. Translate the following into Chinese

(1) It is possible to arrange the list of elements in tabular form with elements having similar properties placed in vertical columns.

(2) Element symbols are abbreviations of the elements' name. In some cases, the abbreviation comes from the element's Latin name. Each symbol is either one or two letters in length.

(3) The percentage of fat varies from person to person, but even an obese person has more water than fat. In a lean male, the percentages of proteins and water are comparable.

(4) Although humans use glucose as an energy source, there is not that much of it free in the bloodstream at any given time.

(5) The groups headed by the members of the two 8-elements are designated as main group elements, and the members of the other groups are called transition or inner transition elements.

2. Translate the following into English

(1) 元素周期表有一个仅包含 2 个元素的非常短的周期，然后是两个包含 8 个元素的短周期，然后是两个包含 18 个元素的长周期。

(2) 元素周期表的任何特定列（组或族）中的元素不仅具有相似的属性，而且当我们在列（族）上下移动时，这些属性也会以一定的规律发生变化。

(3) 这些被称为卤素的元素组成的物质都是有色物质，在表中越往下颜色越深。

(4) 糖和其他碳水化合物仅占人体重的 1% 左右。

(5) 元素周期表如此有用的原因在于它关联了大量有关元素及其化合物的数据，并能对其化学和物理特性进行预测。

UNIT 2 Nomenclature of Inorganic Compounds

Nomenclature is the process of naming chemical compounds with different names so that they can be easily identified as separate chemicals. The living and growing science of chemistry requires a nomenclature system that is expansible and mutable. An international committee has recommended a set of rules for naming compounds, and these are now being adopted throughout the world. One of the principal changes is that proposed by Albert Stock and now known as Stock system for naming compounds of metals (oxides, hydroxides, and salts) in which the metal may exhibit more than one oxidation state. In these cases, the oxidation state of the metal is shown by a Roman numeral in parentheses immediately following the English name of the metal which corresponds to its oxidation number. If the metal has only one common oxidation number, no Roman numeral is used. Another important change is in the naming of complex ions and coordination compounds.

The tendency of atoms to react in ways that achieve an outer shell of eight valence electrons is particularly the main group elements and is given the special name of the octet rule. An atom with almost eight valence electrons tends to gain the needed electrons to have eight electrons in its valence shell and an electron configuration like that of the noble gas nearest to it in atomic number. In gaining electrons, the atom becomes a negatively charged ion called an anion. An atom with only one or two valence electrons tends to lose the number of electrons required to have an electron configuration like the noble gas nearest to it in atomic number. In losing electrons, the atom becomes a positively charged ion called a cation. When an ion forms, the number of protons and neutrons in the nucleus of the atom remains unchanged, only the number of electrons in the valence shell of the atom changes. The scheme is divided into three parts: cations, anions, and compounds (ionic compounds, containing hydrogen, and covalent compounds).

1. Naming Positive Ions (Cations)

A monatomic (containing only one atom) cation forms when a metal loses one or more valence electrons. Monatomic cations should be named as the corresponding element, without change or suffix. Indication of the electrochemical valency in the names of compounds should be made only by Stock's method. This is done by means of Roman figures, placed in parentheses and following, without a hyphen, immediately after the names of the elements they refer to. The system of valency indication by terminations, such as -ous, -ic (ferrous, ferric), was previously used (Fig.1.5).

A polyatomic ion contains more than one atom. Names for polyatomic cations derived by addition of protons to monatomic anions are formed by adding the ending -onium to the root of the name of the anion element. For example: phosphonium, oxonium, sulfonium, selenonium, telluronium, and iodonium ions.

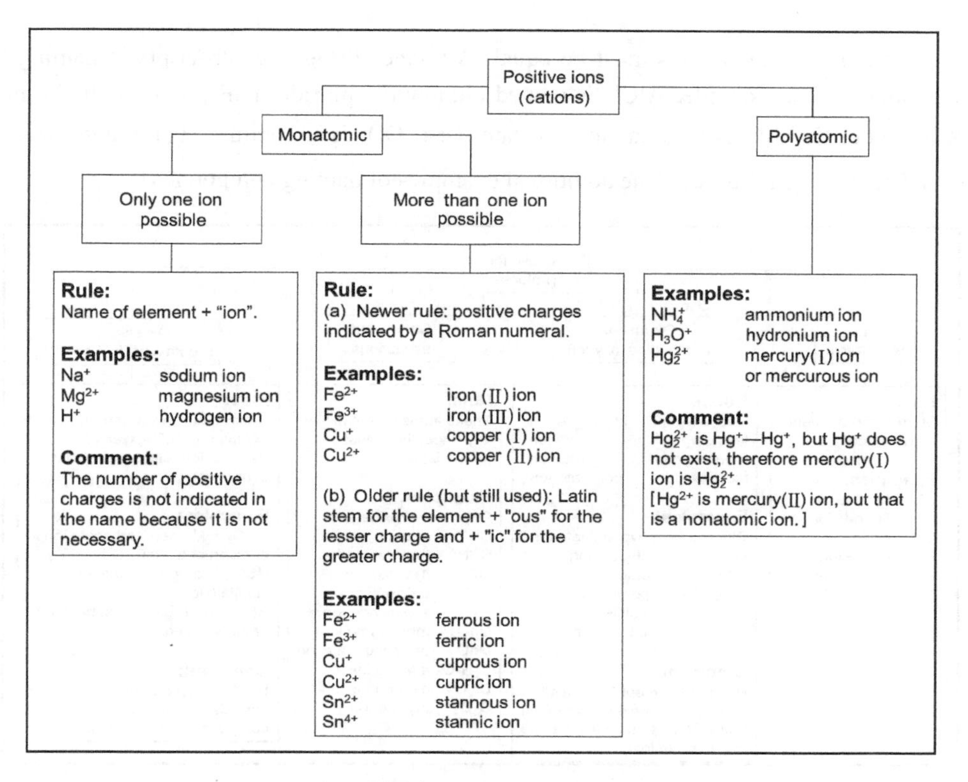

Fig.1.5 Scheme for naming cations

2. Naming Negative Ions (Anions)

Negative ions, anions, may be monatomic or polyatomic. The names for monatomic anions shall consist of the element's name (sometimes abbreviated), with the termination -ide. Two polyatomic anions which also have names ending with -ide are the hydroxide ion, OH^-, and the cyanide ion, CN^-.

Many polyatomic anions contain oxygen in addition to another element. The number of oxygen atoms in such oxyanions is denoted by the use of the suffixes -ite and -ate, meaning fewer and more oxygen atoms, respectively. In cases where it is necessary to denoted more than two oxyanions of the same element, the prefixes hypo- and per-, meaning still fewer and still more oxygen atoms, respectively, may be used. A series of anions is named in Fig.1.6.

3. Naming Compounds (Ionic, containing hydrogen, and covalent compounds)

Many chemical compounds are essentially binary in nature and can be regarded as combinations of ions or radicals, others may be treated as such for the purpose of nomenclature. The name of the compound consists of the name of the metal from which the cation (positive ion) was formed, followed by the name of the anion (negative ion). Although ionic compounds do not consist of molecules, they do have a definite ratio of one kind of ion to another, their formulas give this ratio. The formula of an ionic compound shows the simplest whole-number ratio between cations and anions. For example, NaCl represents the simplest ratio of sodium ions to chloride ions, namely 1 : 1. In an ionic compound, the total number of positive charges of the cations and the total number of negative charges of the anions must be equal. We generally ignore subscripts in naming binary ionic compounds. For example, $AlCl_3$ is named aluminum chloride, $FeBr_2$ is iron (II) bromide or ferrous bromide, $Ca(C_2H_3O_2)_2$ is calcium acetate, $Cr_2(SO_4)_3$ is chromium (III) sulfate or chromic sulfate, and so on. Fig.1.7 gives some additional examples of naming compounds.

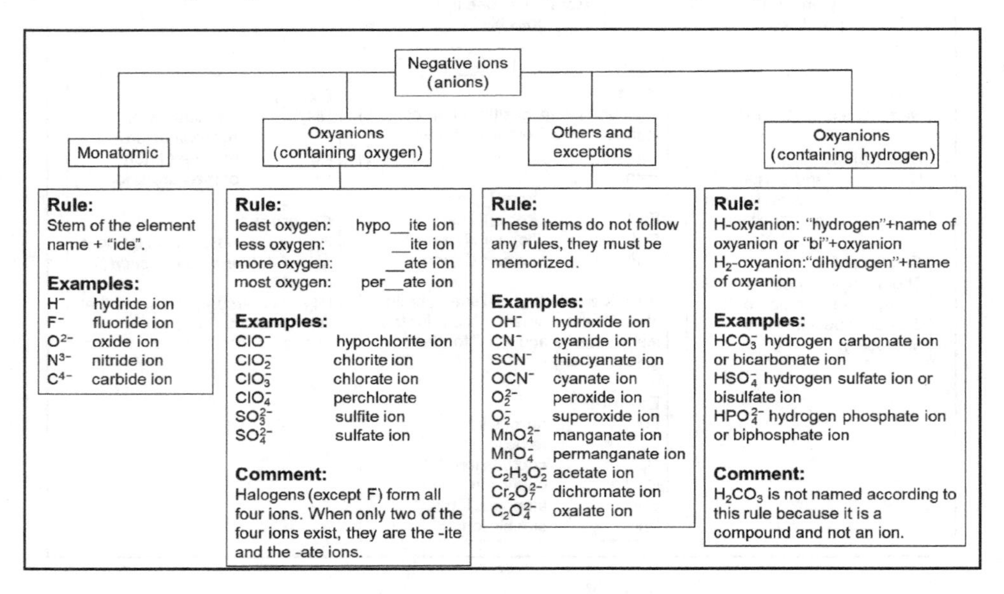

Fig.1.6 Scheme for naming anions

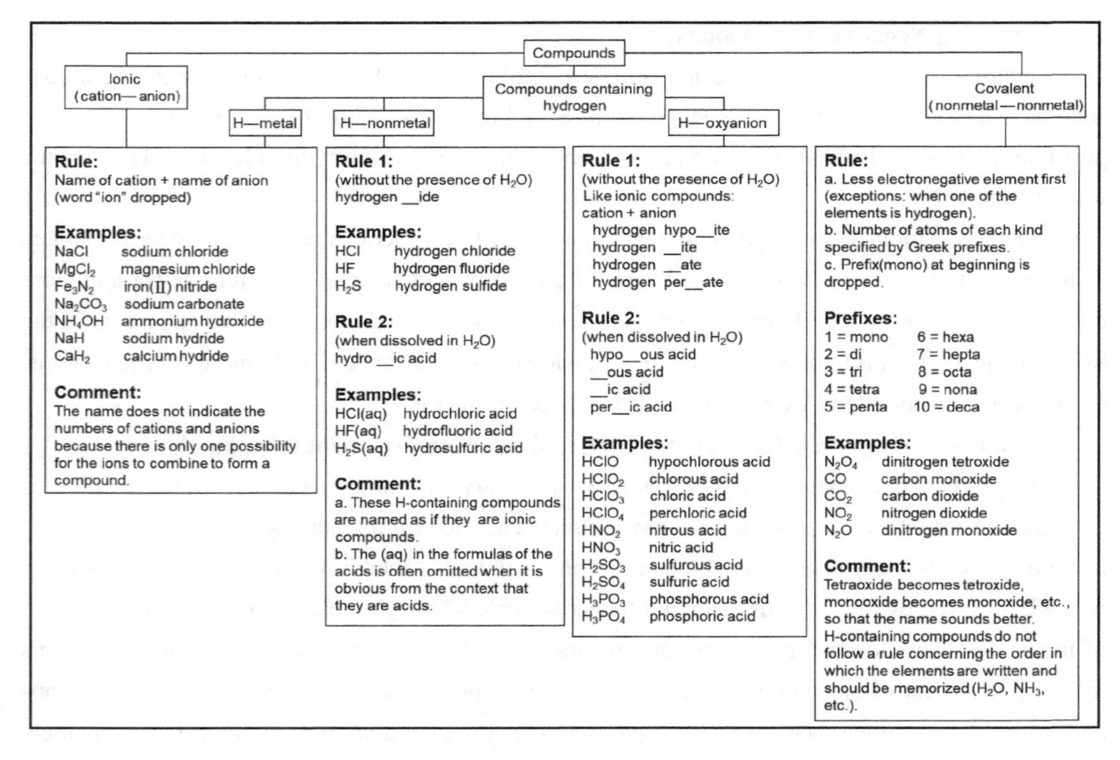

Fig.1.7 Scheme for naming compounds

Acid names may be obtained directly by changing the name of acid ion (negative ion) in Fig.1.7. There are few cases where the name of acid is changed slightly from that of the acid radical. For example, H_2SO_4, is sulfuric acid rather than sulfic. Similarly, H_3PO_4, is phosphoric acid rather than phosphic.

The older system of naming, one still widely used, employs Greek prefixes for both the number of oxygen atoms and that of the other element in the compounds. The prefixes used are ① mono-, sometimes reduced to mon-, ②di-, ③tri-, ④tetra-, ⑤penta-, ⑥hexa-, ⑦hepta-, ⑧octa-, ⑨nona- and ⑩deca-. Generally, the letter a is omitted from the prefix (from tetra on) when naming a nonmetal oxide and often mono- is omitted from the name altogether.

4. Acid Salts

It is conceivable that in the neutralization of an acid by a base, only a part of the hydrogen might be neutralized, thus

$$NaOH+H_2SO_4 \longrightarrow NaHSO_4+H_2O$$

The compound $NaHSO_4$ has acid properties since it contains hydrogen, and is also a salt since it contains both a metal and an acid radical. Such a salt containing acidic hydrogen is termed an acid salt. Phosphoric acid (H_3PO_4) might be progressively neutralized to form the salts: NaH_2PO_4, Na_2HPO_4, and Na_3PO_4. The first two are acid salts, since they contain replaceable hydrogen. A way of naming these salts is to call Na_2HPO_4— disodium hydrogen phosphate and NaH_2PO_4— sodium dihydrogen phosphate. These acid phosphates are important in controlling the alkalinity of blood. The third compound, sodium phosphate Na_3PO_4, which contains no replaceable hydrogen, is often referred to as normal sodium phosphate, or trisodium phosphate to differentiate it from the two acid salts.

Words and Expressions

nomenclature *n.* 命名法；术语
cation *n.* 阳离子；正离子
anion *n.* 阴离子；负离子
compound *n.* 化合物；混合物；复合词
adj. 复合的；混合的
v. 合成；混合
oxide *n.* 氧化物
hydroxide *n.* 氢氧化物；羟化物
hydroxide ion 氢氧根离子
hydroxyl group 羟基（基团）
oxyanion *n.* 氧离子
covalent *adj.* 共价的；共有原子价的

valency *n.* 价；原子价；化合价
iron (Ⅱ) bromide 溴化亚铁
ferrous *adj.* 铁的；含铁的；亚铁的
ferrous bromide 溴化亚铁
ferric *adj.* 铁的；含铁的；三价铁的
stannous *adj.* 亚锡的；二价锡的
stannic *adj.* 锡的；正锡的；四价锡的
mercurous *adj.* 水银的；亚汞的；一价汞的
mercuric *adj.* 汞的；正汞的；二价汞的
cupric *adj.* 正铜的；二价铜的
monatomic *adj.* 单原子的

polyatomic *adj.* 多原子的；多元的

phosphoric *adj.* 含磷的

phosphoric acid 磷酸

hypochlorous acid 次氯酸

chlorous acid 亚氯酸

chloric acid 氯酸

perchloric acid 高氯酸

dinitrogen tetroxide 四氧化二氮

nitrogen dioxide 二氧化氮

dinitrogen monoxide 一氧化二氮

Notes

➢ In these cases, the oxidation state of the metal is shown by a Roman numeral in parentheses immediately following the English name of the metal which corresponds to its oxidation number.

参考译文：在这些情况下，金属的氧化态用名称后的括号中的罗马数字表示，该数字与金属的氧化数一致。

➢ Monatomic cations should be named as the corresponding element, without change or suffix.

参考译文：单原子的阳离子用其对应的元素命名，不用进行修改或者加后缀。

➢ The name of the compound consists of the name of the metal from which the cation (positive ion) was formed, followed by the name of the anion (negative ion).

参考译文：化合物的名称由金属阳离子(正离子)的名称加上阴离子(负离子)的名称组成。

➢ In an ionic compound, the total number of positive charges of the cations and the total number of negative charges of the anions must be equal.

参考译文：在离子化合物中，阳离子的正电荷总数和阴离子的负电荷总数必须相等。

Exercises

1. Give the IUPAC names of the following

(1) $NaOH$ (2) Hg_2Cl_2 (3) $Hg(ClO_4)_2$ (4) Na_2CO_3 (5) $NaHSO_3$ (6) $(NH_4)_2SO_4$

2. Draw structural formulas of the following

(1) hydrogen peroxide

(2) sulfate

(3) potassium permanganate

(4) chromium trioxide

(5) sodium potassium sulfate

(6) hydrocyanic acid

3. Translate the following into Chinese

(1) Negative ions, anions, may be monatomic or polyatomic.

(2) Many polyatomic anions contain oxygen in addition to another element. The number of oxygen atoms in such oxyanions is denoted by the use of the suffixes -ite and -ate, meaning fewer and more oxygen atoms, respectively.

(3) In cases where it is necessary to denote more than two oxyanions of the same element, the prefixes hypo- and per-, meaning still fewer and still more oxygen atoms, respectively, may be used.

(4) The compound $NaHSO_4$ has acid properties since it contains hydrogen, and is also a salt

since it contains both a metal and an acid radical. Such a salt containing acidic hydrogen is termed an acid salt.

(5) The first two are acid salts, since they contain replaceable hydrogen. A way of naming these salts is to call Na_2HPO_4—disodium hydrogen phosphate and NaH_2PO_4—sodium dihydrogen phosphate. These acid phosphates are important in controlling the alkalinity of blood. The third compound, sodium phosphate Na_3PO_4, which contains no replaceable hydrogen, is often referred to as normal sodium phosphate, or trisodium phosphate to differentiate it from the two acid salts.

UNIT 3 Nomenclature of Organic Compounds

1. Organic Chemistry

The modern definition of organic chemistry is the chemistry of carbon compounds. What is so special about carbon that a whole branch of chemistry is devoted to its compounds? Unlike most other elements, carbon forms strong bonds to other carbon atoms and a wide variety of other elements. Chains and rings of carbon atoms can be built up to form an endless variety of molecules. It is this diversity of carbon compounds that provides the basis for life on earth. Living creatures are composed largely of complex organic compounds that serve structural, chemical, or genetic functions.

In the nineteenth century, experiments showed that organic compounds could be synthesized from inorganic compounds. In 1828, the German chemist Friedrich Wöhler converted ammonium cyanate, made from ammonia and cyanic acid, to urea simply by heating it in the absence of oxygen.

For the purpose of this brief survey, we divide organic compounds into three major classes: ①hydrocarbons, ②compounds containing oxygen, ③compounds containing nitrogen (Table 1.1).

Table 1.1 Six common fuctional groups of organic compounds

Family	Functional Group	Example	Name
Alcohol	—OH	CH_3CH_2OH	Ethanol
Amine	$—NH_2$	$CH_3CH_2NH_2$	Ethylamine
Aldehyde	$-\overset{\overset{O}{\|}}{C}-H$	$CH_3\overset{\overset{O}{\|}}{C}H$	Acetaldehyde
Ketone	$-\overset{\overset{O}{\|}}{C}-$	$CH_3\overset{\overset{O}{\|}}{C}CH_3$	Acetone
Carboxylic acid	$-\overset{\overset{O}{\|}}{C}-OH$	$CH_3\overset{\overset{O}{\|}}{C}OH$	Acetic acid
Ester	$-\overset{\overset{O}{\|}}{C}-OR$	$CH_3\overset{\overset{O}{\|}}{C}OCH_2CH_3$	Ethyl acetate

Hydrocarbons are compounds composed entirely of carbon and hydrogen. The major classes of hydrocarbons are alkanes, alkenes, alkynes, and aromatic hydrocarbons (Fig.1.8).

2. Hydrocarbons

(1) Alkanes

Alkanes are hydrocarbons that contain only single bonds. Alkane names generally have the -ane suffix, and the first part of the name indicates the number of carbon atoms. Table 1.2 shows how the prefixes in the names correspond with the number of carbon atoms.

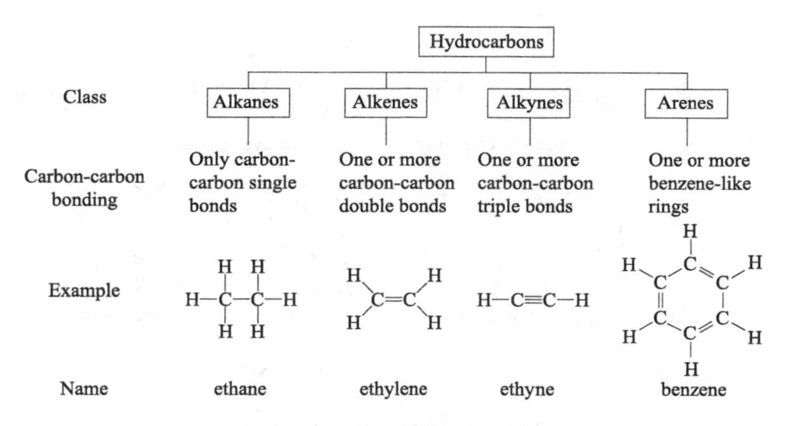

Fig.1.8 Major classes of hydrocarbons

Table 1.2 Correspondence of prefixes and numbers of carbon atoms

Alkane Name	Number of Carbons	Alkane Name	Number of Carbons
Methane	1	Hexane	6
Ethane	2	Heptane	7
Propane	3	Octane	8
Butane	4	Nonane	9
Pentane	5	Decane	10

CH_4	$CH_3—CH_3$	$CH_3—CH_2—CH_3$	$CH_3—CH_2—CH_2—CH_3$	$CH_3—CH(CH_3)—CH_3$
Methane	Ethane	Propane	Butane	Isobutane

Alkanes are the major components of heating gases (natural gas and liquefied petroleum gas), gasoline, jet fuel, diesel fuel, motor oil, fuel oil, and paraffin "wax". Other than combustion, alkanes undergo few reactions. In fact, when a molecule contains an alkane portion and a non-alkane portion, we often ignore the presence of the alkane portion because it is relatively unreactive. Alkanes undergo few reactions because they have no functional group, the part of the molecule where reactions usually occur. Functional groups are distinct chemical units, such as double bonds, hydroxyl groups, or halogen atoms, that are reactive. Most organic compounds are characterized and classified by their functional groups.

An alkyl group is an alkane portion of a molecule, with one hydrogen atom removed to allow bonding to the rest of the molecule. An ethyl group (C_2H_5) is attached to cyclohexane to give ethylcyclohexane. We might try to name this compound as "cyclohexylethane", but we should treat the larger fragment as the parent compound (cyclohexane), and the smaller group as the alkyl group (ethyl).

The cycloalkanes are a special class of alkanes in the form of a ring (Fig.1.9). With the Lewis structures and line-angle formulas of cyclopentane and cyclohexane, these cycloalkanes contain five and six carbons, respectively.

In these cases, we can use the symbol R as a substituent to represent an alkyl group (or some other unreactive group).

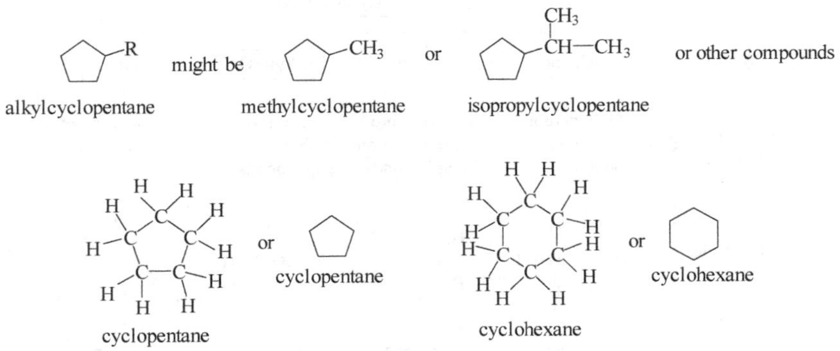

alkylcyclopentane might be methylcyclopentane ($-CH_3$) or isopropylcyclopentane ($-CH(CH_3)-CH_3$) or other compounds

cyclopentane or cyclopentane

cyclohexane or cyclohexane

Fig.1.9 Chemical structures of cycloalkanes

Most alkanes have structural isomers, and we need a way of naming all the different isomers. For example, there are two isomers of formula C_4H_{10}. The unbranched isomer is simply called butane (or *n*-butane, meaning "normal" butane), and the branched isomer is called isobutane, meaning an "isomer of butane". The three isomers of C_5H_{12} are called pentane (or *n*-pentane), isopentane, and neopentane (Fig.1.10).

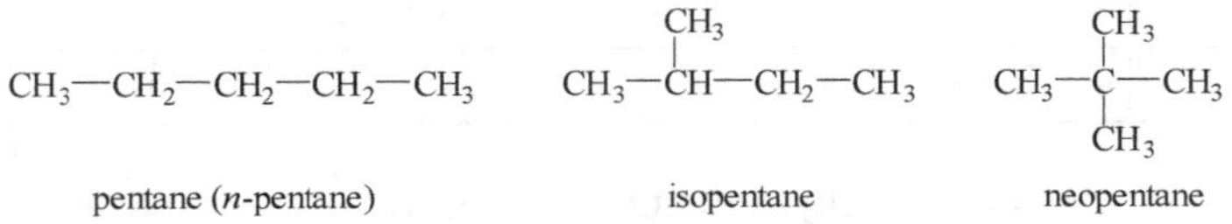

$CH_3-CH_2-CH_2-CH_2-CH_3$ pentane (*n*-pentane)

$CH_3-CH(CH_3)-CH_2-CH_3$ isopentane

$CH_3-C(CH_3)_2-CH_3$ neopentane

Fig.1.10 The three isomers of C_5H_{12}

The IUPAC rules are accepted throughout the world as the standard method for naming organic compounds. The names that are generated using this system are called IUPAC names or systematic names. The IUPAC system uses the longest chain of carbon atoms as the main chain, which is numbered to give the locations of side chains. Four rules govern this process.

RULE 1: Finding the Main Chain

The first rule of nomenclature gives the base name of the compound. Find the longest continuous chain of carbon atoms, and use the name of this chain as the base name of the compound. For example, the longest chain of carbon atoms in the compound in the margin contains six carbons, so the compound is named as hexane derivative. The longest chain is rarely drawn in a straight line, look carefully to find it. The following compound contains two different seven-carbon chains and is named as heptane. We choose the chain on the right as the main chain because it has more substituents attached to the chain (Fig.1.11).

RULE 2: Numbering the Main Chain

To give the locations of the substituents, assign a number to each carbon atom on the main chain. Number the longest chain, beginning with the end of the chain nearest a substituent.

We start the numbering from the end nearest a branch so the numbers of the substituted carbons will be as low as possible. In the preceding heptane structure on the right (Fig.1.11), numbering from top to bottom gives the first branch at C3 (carbon atom 3), but numbering from bottom to top gives the first branch at C2 (Fig.1.12). Numbering from bottom to top is correct. (If each end had a substituent in the same distance, we would start at the end nearer the second branch point.)

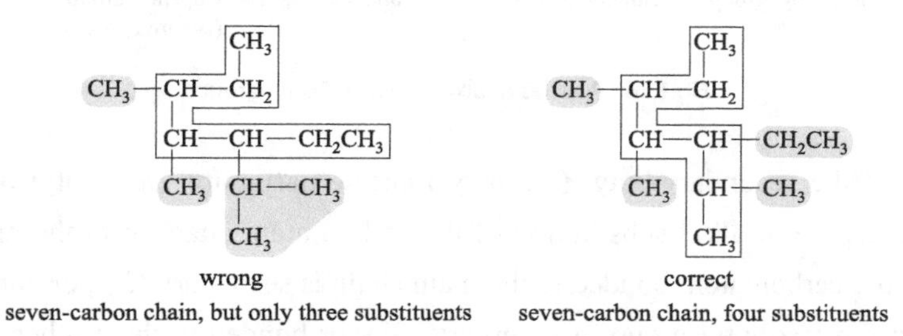

Fig.1.11 Finding the main chain of hexane derivatives

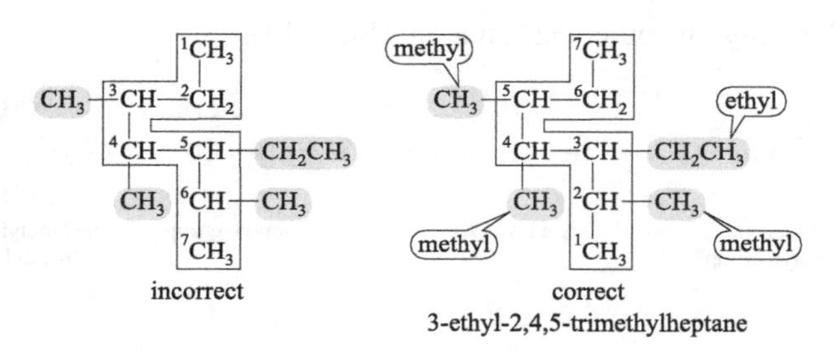

Fig.1.12 Numbering the main chain of hexane derivatives

RULE 3: Naming Alkyl Groups

Name the substituent groups attached to the longest chain as alkyl groups. Give the location of each alkyl group by the number of the main-chain carbon atom to which it is attached.

Alkyl groups are named by replacing the -ane suffix of the alkane name with -yl (Fig.1.13). Methane becomes methyl, ethane becomes ethyl. You may encounter the word amyl, which is an archaic term for a pentyl (five-carbon) group.

CH_4	methane	CH_3-	methyl group
CH_3-CH_3	ethane	CH_3-CH_2-	ethyl group
$CH_3-CH_2-CH_3$	propane	$CH_3-CH_2-CH_2-$	propyl group
$CH_3-(CH_2)_2-CH_3$	butane	$CH_3-(CH_2)_2-CH_2-$	butyl group
$CH_3-(CH_2)_3-CH_3$	pentane	$CH_3-(CH_2)_3-CH_2-$	pentyl group (*n*-amyl group)

Fig.1.13 Chemical structures of alkanes with unbranched chains

The propyl and butyl groups are simply unbranched three-carbon and four-carbon alkyl groups. These groups are sometimes named as "*n*-propyl" and "*n*-butyl" groups, to distinguish them from other kinds of (branched) propyl and butyl groups. The simple branched alkyl groups are usually known by common names. The isopropyl and isobutyl groups have a characteristic "iso" grouping, just as in isobutane (Fig.1.14).

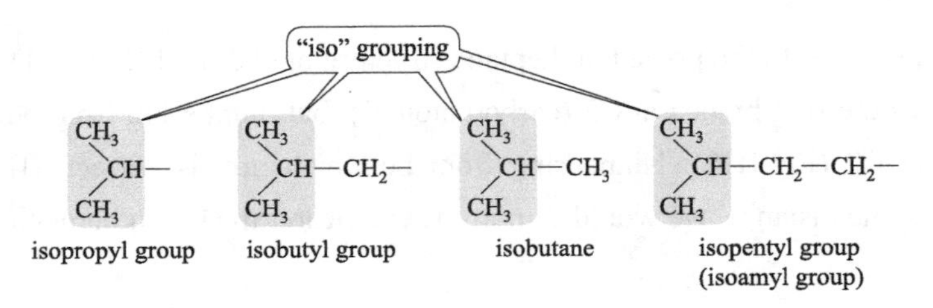

Fig.1.14 Names of alkanes with "iso" groups

The names of the secondary-butyl (*sec*-butyl) and tertiary-butyl (*tert*-butyl or *t*-butyl) groups are based on the degree of alkyl substitution of the carbon atom attached to the main chain. In the *sec*-butyl group, the carbon atom bonded to the main chain is secondary (2°), or bonded to two other carbon atoms. In the *tert*-butyl group, it is tertiary (3°), or bonded to three other carbon atoms. In both the *n*-butyl group and the isobutyl group, the carbon atoms bonded to the main chain are primary (1°), bonded to only one other carbon atom (Fig.1.15).

$CH_3—CH_2—CH_2—CH_2—$ butyl group (or "*n*-butyl group")

$CH_3—CH(CH_3)—CH_2—$ isobutyl group

$CH_3—CH_2—CH(CH_3)—$ *sec*-butyl group

$CH_3—C(CH_3)_2—$ *tert*-butyl group (or "*t*-butyl group")

$R—CH_2—$ a primary (1°) carbon

$R—CH(R)—$ a secondary (2°) carbon

$R—C(R)_2—$ a tertiary (3°) carbon

$CH_3CH_2CH_2—CH_2—$ *n*-butyl group (1°)

$CH_3CH_2—CH(CH_3)—$ *sec*-butyl group (2°)

$CH_3—C(CH_3)_2—$ *tert*-butyl group (3°)

Fig.1.15 Names of alkanes with butyl groups

Haloalkanes can be named just like alkanes, with the halogen atom treated as a substituent (Fig.1.16). Halogen substituents are named fluoro-, chloro-, bromo-, and iodo-.

$CH_3—CH(Br)—CH_2CH_3$ 2-bromobutane

$CH_3—CH(CH_3)—CH(Cl)—CH_2CH_3$ 3-chloro-2-methylpentane

$CH_3—CH(F)—CH_2F$ 1,2-difluoropropane

Fig.1.16 Names of haloalkanes

RULE 4: Organizing Multiple Groups

The final rule deals with naming compounds with more than one substituent. When two or more substituents are present, list them in alphabetical order. When two or more of the same alkyl substituents are present, use the prefixes di-, tri-, tetra-, etc. to avoid having to name the alkyl group twice (di- means 2; tri- means 3; tetra- means 4; penta- means 5; hexa- means 6; hepta-

means 7). Include a position number for each substituent, even if it means repeating a number more than once.

(2) Alkenes

Alkenes are hydrocarbons that contain carbon-carbon double bonds. Alkene names end in the -ene suffix. Simple alkenes are named much like alkanes, using the root name of the longest chain containing the double bond. The ending is changed from -ane to -ene. For example, "ethane" becomes "ethene", "propane" becomes "propene" and "cyclohexane" becomes "cyclohexene".

In 1993, the IUPAC recommended a logical change in the positions of the numbers used in names. Instead of placing the numbers before the root name (1-butene), they recommended placing them immediately before the part of the name they locate (but-1-ene) (Fig.1.17).

	$CH_2{=}CH{-}CH_2{-}CH_3$ (C1–C4)	$CH_2{=}CH{-}CH_2{-}CH_2{-}CH_3$ (C1–C5)	
old IUPAC names:	1-butene	1-pentene	
new IUPAC names:	but-1-ene	pent-1-ene	
	$CH_3{-}CH{=}CH{-}CH_3$ (C1–C4)	$CH_3{-}CH{=}CH{-}CH_2{-}CH_3$ (C1–C5)	
old IUPAC names:	2-butene	2-pentene	
new IUPAC names:	but-2-ene	pent-2-ene	
	$CH_3{-}CH{=}C(CH_3){-}CH_3$ (C4–C1)	$CH_3{-}CH(CH_3){-}CH{=}CH_2$ (C4–C1)	$CH_3{-}CH{=}C(CH_3){-}CH_2{-}CH_2{-}CH(CH_3){-}CH_3$ (C1–C7)
old IUPAC names:	2-methyl-2-butene	3-methyl-1-butene	3,6-dimethyl-2-heptene
new IUPAC names:	2-methylbut-2-ene	3-methylbut-1-ene	3,6-dimethylhept-2-ene

3-propyl-1-heptene
3-propylhept-1-ene

Fig.1.17 Names of alkenes

Each alkyl group attached to the main chain is listed with a number to give its location. Note that the double bond is still given preference in numbering, however.

Cycloalkenes are assumed to have the double bond in the number 1 position (Fig.1.18).

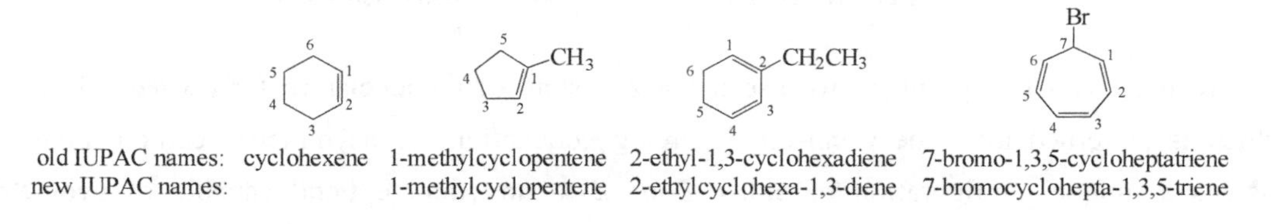

Fig.1.18 Names of cycloalkenes

The rigidity and lack of rotation of carbon-carbon double bonds give rise to *cis-trans* isomerism, also called geometric isomerism. If two similar groups bonded to the carbons of the double bond are on the same side of the bond, the alkene is *cis* isomer. If the similar groups are on opposite sides of the bond, the alkene is *trans* isomer. Not all alkenes are capable of showing *cis-trans* isomerism. If either carbon of the double bond holds two identical groups, the molecule cannot have *cis* and *trans* forms. Following are some *cis* and *trans* alkenes and some alkenes that cannot show *cis-trans* isomerism (Fig.1.19).

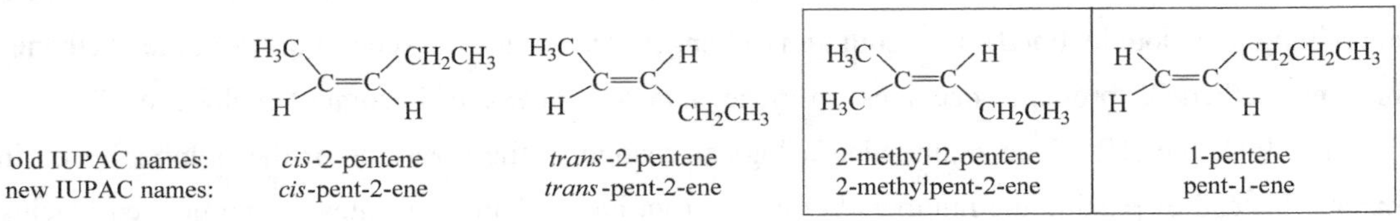

old IUPAC names:	*cis*-2-pentene	*trans*-2-pentene	2-methyl-2-pentene	1-pentene
new IUPAC names:	*cis*-pent-2-ene	*trans*-pent-2-ene	2-methylpent-2-ene	pent-1-ene

Fig.1.19 Names of *cis-trans* alkene isomerism

Trans cycloalkenes are unstable unless the ring is large enough (at least eight carbon atoms) to accommodate the *trans* double bond. Therefore, all cycloalkenes are assumed to be *cis* unless they are specifically named *trans*. The *cis* name is rarely used with cycloalkenes, except to distinguish a large cycloalkene from its *trans* isomer (Fig.1.20).

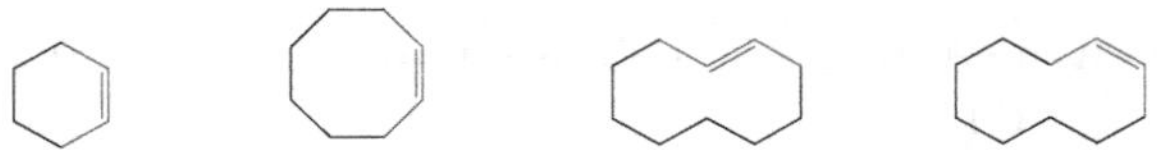

cyclohexene cyclooctene *trans*-cyclodecene *cis*-cyclodecene

Fig.1.20 Names of cycloalkenes

The *cis-trans* nomenclature for geometric isomers sometimes gives an ambiguous name. For example, the isomers of 1-bromo-1-chloropropene are not clearly *cis* or *trans* because it is not obvious which substituents are referred to as *cis* or *trans* (Fig.1.21).

geometric isomers of 1-bromo-1-chloropropene

Fig.1.21 The *cis-trans* nomenclature of geometric isomers

To deal with this problem, we use the *E-Z* system of nomenclature for *cis-trans* isomers, which is patterned after the Cahn-Ingold-Prelog convention for asymmetric carbon atoms. It assigns a unique configuration of either *E* or *Z* to any double bond capable of geometric isomerism.

To name an alkene by the *E-Z* system, mentally separate the double bond into its two ends. Consider each end of the double bond separately, and use those same rules to assign first and second priorities to the two substituent groups on that end. Do the same for the other end of the double bond. If the two first-priority atoms are together (*cis*) on the same side of the double bond, you have the *Z* isomer, from the German word zusammen, "together". If the two first-priority atoms are on opposite (*trans*) sides of the double bond, you have the *E* isomer, from the German word entgegen, "opposite" (Fig.1.22).

zusammen　　entgegen

Fig.1.22　The *E*–*Z* isomers of alkenes

For example (Fig.1.23)

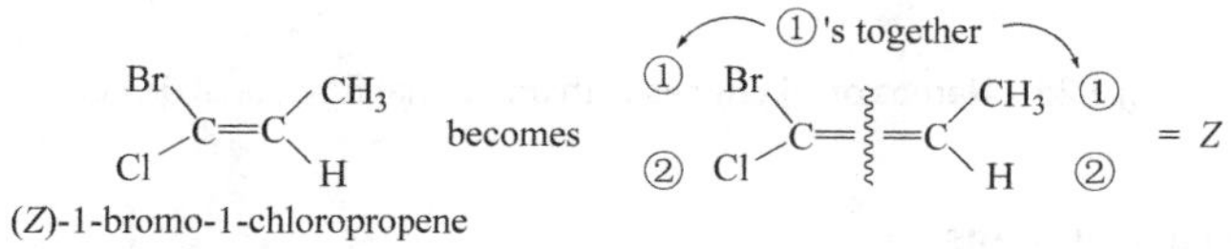

(*Z*)-1-bromo-1-chloropropene　becomes　①'s together　= *Z*

Fig.1.23　The *E*–*Z* isomers of haloalkenes

(3) Alkynes

Alkynes are hydrocarbons with carbon-carbon triple bonds as their functional group. Alkyne names generally have the -yne suffix, although some of their common names (for example, acetylene) do not conform to this rule (Fig.1.24). The triple bond is linear, so there is no possibility of geometric (*cis-trans*) isomerism in alkynes.

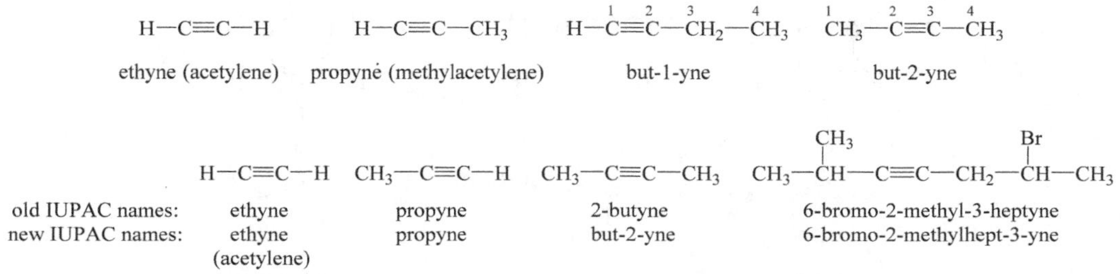

$H-C\equiv C-H$	$H-C\equiv C-CH_3$	$H-C\equiv C-CH_2-CH_3$	$CH_3-C\equiv C-CH_3$
ethyne (acetylene)	propyne (methylacetylene)	but-1-yne	but-2-yne

	$H-C\equiv C-H$	$CH_3-C\equiv C-H$	$CH_3-C\equiv C-CH_3$	$CH_3-CH(CH_3)-C\equiv C-CH_2-CH(Br)-CH_3$
old IUPAC names:	ethyne	propyne	2-butyne	6-bromo-2-methyl-3-heptyne
new IUPAC names:	ethyne (acetylene)	propyne	but-2-yne	6-bromo-2-methylhept-3-yne

Fig.1.24　Names of alkynes

Alkynes are similar to that for alkenes. In an alkyne, four atoms must be in a straight line. These four collinear atoms are not easily bent into a ring, so cycloalkynes are rare. Cycloalkynes are stable only if the ring is large, containing eight or more carbon atoms. We find the longest continuous chain of carbon atoms that includes the triple bond and change the -ane ending of the parent alkane to -yne. The chain is numbered from the end closest to the triple bond, and the position of the triple bond is designated by its lower-numbered carbon atom. Substituents are given numbers to indicate their locations.

When additional functional groups are present, the suffixes are combined to produce the compound names of the alkenynes (a double bond and a triple bond), alkynols (a triple bond and an alcohol), and so on. Most alkynes can be named as a molecule of acetylene with one or two alkyl substituents. This nomenclature is like the common nomenclature for ethers, where we name the two alkyl groups bonded to oxygen (Fig.1.25).

$H—C{\equiv}C—H$ acetylene	$R—C{\equiv}C—H$ alkylacetylene	$R—C{\equiv}C—R'$ dialkylacetylene
$CH_3—C{\equiv}C—H$ methylacetylene	$Ph—C{\equiv}C—H$ phenylacetylene	$CH_3—C{\equiv}C—CH_2CH_3$ ethylmethylacetylene
$(CH_3)_2CH—C{\equiv}C—CH(CH_3)_2$ diisopropylacetylene	$Ph—C{\equiv}C—Ph$ diphenylacetylene	$H—C{\equiv}C—CH_2OH$ hydroxymethylacetylene (propargyl alcohol)

Fig.1.25 Names of alkenynes with additional functional groups

(4) Aromatic Hydrocarbons

The following compounds may look like cycloalkenes, but their properties are different from those of simple alkenes. These aromatic hydrocarbons (also called arenes) are all derivatives of benzene, represented by a six-membered ring with three double bonds (Fig.1.26). This bonding arrangement is particularly stable. Just as a generic alkyl group substituent is represented by R, a generic aryl group is represented by Ar. When a benzene ring serves as a substituent, it is called a phenyl group, abbreviated Ph.

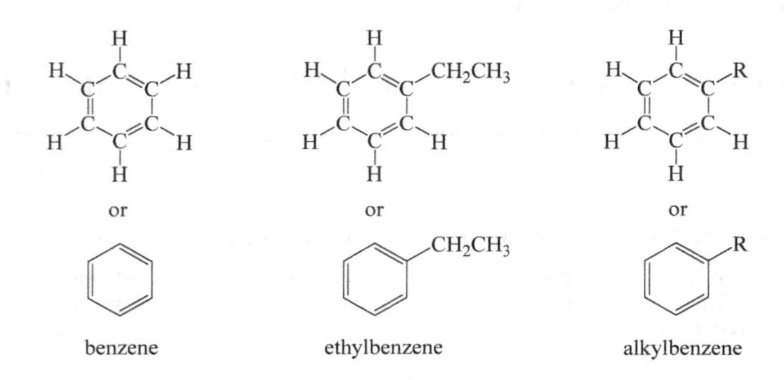

Fig.1.26 Names of arenes

Some ordinary aromatic hydrocarbons, such as benzene, toluene, xylene and naphthalene, are mainly obtained from coal and petroleum. They are raw materials for preparation of other aromatic compounds. The generic name of aromatic hydrocarbons is “arene”.

The monocyclic substituted aromatic hydrocarbons are named as derivatives of benzene. The position of substituents is indicated by numbering the carbon atoms of the benzene ring clockwise or counter-clockwise. The lowest numbers possible are given to substituents. When two substituents are present, *o*- (*ortho*), *m*- (*meta*) and *p*- (*para*) may be used instead of “1, 2-” “1, 3-” and “1, 4-” respectively (Fig.1.27).

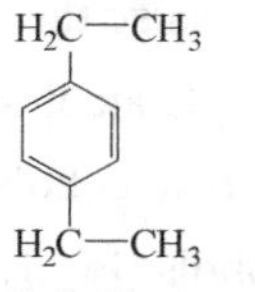

H_2C-CH_3 / H_2C-CH_3 H_2C-CH_3, CH_3, CH_3

p-diethylbenzene 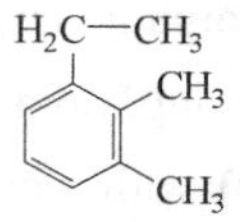   1, 2-dimethyl-3-ethylbenzene

Fig.1.27 Names of monocyclic substituted aromatic hydrocarbons

Some customary or trivial names of aromatic compounds are retained (Fig.1.28).

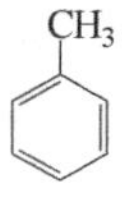

CH_3 CH_3, CH_3 $H_3C-CH-CH_3$ $HC=CH_2$

toluene 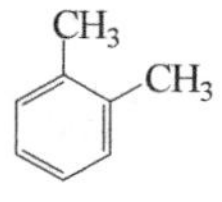   xylene 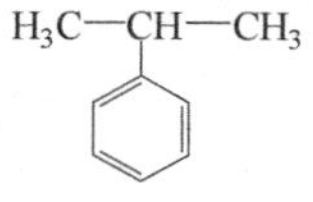   cumene 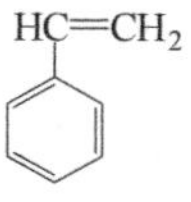   styrene

Fig.1.28 Customary names of aromatic compounds

Besides the monocyclic aromatic hydrocarbons, there are many other aromatic ring systems, called condensed or fused rings. The following are some examples (Fig.1.29).

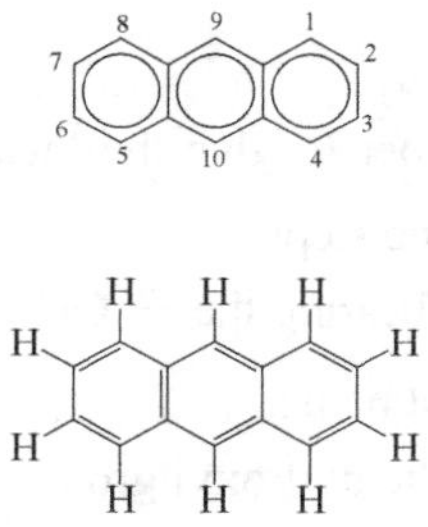

anthracene 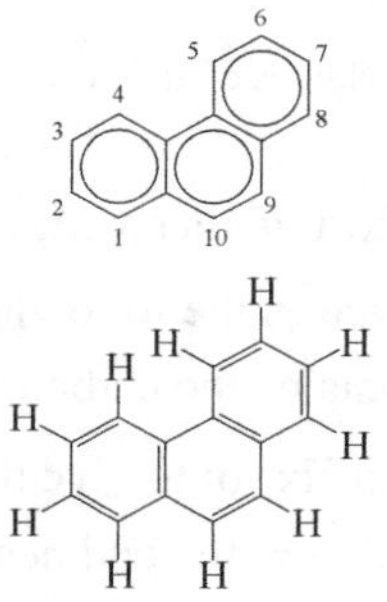   phenanthrene 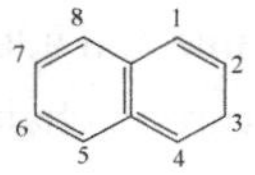   naphthalene 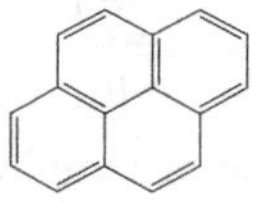   pyrene

Fig.1.29 Names of aromatic hydrocarbons with condensed or fused rings

Heterocyclic compounds, with rings containing sp^2 hybridized atoms of other elements, can also be aromatic. Nitrogen, oxygen, and sulfur are the most common heteroatoms in heterocyclic aromatic compounds. The polynuclear aromatic hydrocarbons (abbreviated PAHs or PNAs) are composed of two or more fused benzene rings. Fused rings share two carbon atoms and the bond between them (Fig.1.30).

3. Organic Compounds Containing Oxygen

Many organic compounds contain oxygen atoms bonded to alkyl groups. The major classes of oxygen-containing compounds are alcohols, ethers, ketones, aldehydes, carboxylic acids, and acid derivatives.

(1) Alcohols

Alcohols are organic compounds that contain the hydroxyl group as their functional group.

Alcohols are among the most polar organic compounds because the hydroxyl group is strongly polar and can participate in hydrogen bonding. Some of the simple alcohols like ethanol and methanol are miscible (soluble in all proportions) with water. Names of alcohols end in the -ol suffix from the word "alcohol", as shown for the following common alcohols (Fig.1.31).

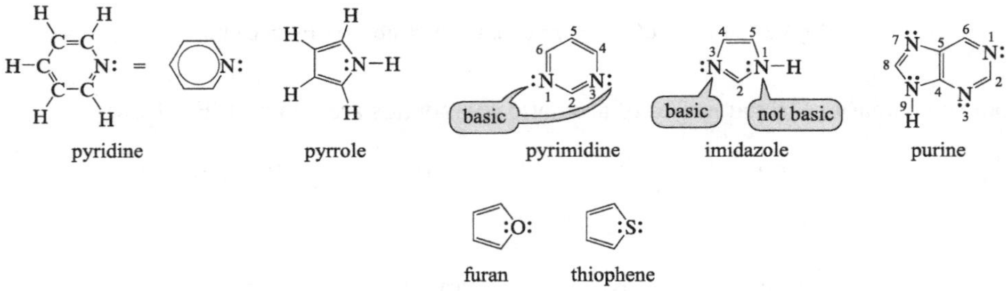

pyridine pyrrole pyrimidine imidazole purine

furan thiophene

Fig.1.30 Names of heterocyclic aromatic hydrocarbons

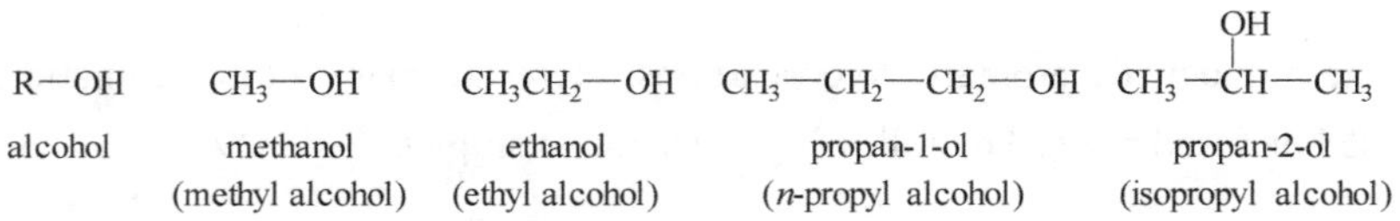

R—OH	CH_3—OH	CH_3CH_2—OH	CH_3—CH_2—CH_2—OH	CH_3—CH(OH)—CH_3
alcohol	methanol (methyl alcohol)	ethanol (ethyl alcohol)	propan-1-ol (*n*-propyl alcohol)	propan-2-ol (isopropyl alcohol)

Fig.1.31 Names of alcohols

In general, the name carries the -ol suffix, together with a number to give the location of the hydroxyl group. The formal rules are summarized in the following three steps:

① Name the longest carbon chain that contains the carbon atom bearing the —OH group. Drop the final -e from the alkane name and add the suffix -ol to give the root name.

② Number the longest carbon chain starting at the end nearest the hydroxyl group, and use the appropriate number to indicate the position of the —OH group. (The hydroxyl group takes precedence over double and triple bonds.)

③ Name all the substituents and give their numbers, as you would for an alkane or an alkene.

(2) Phenols

Because the phenol structure involves a benzene ring, the terms *ortho* (1,2-disubstituted), *meta* (1,3-disubstituted), and *para* (1,4-disubstituted) are often used in the common names. The following examples illustrate the systematic names and the common names of some simple phenols (Fig.1.32).

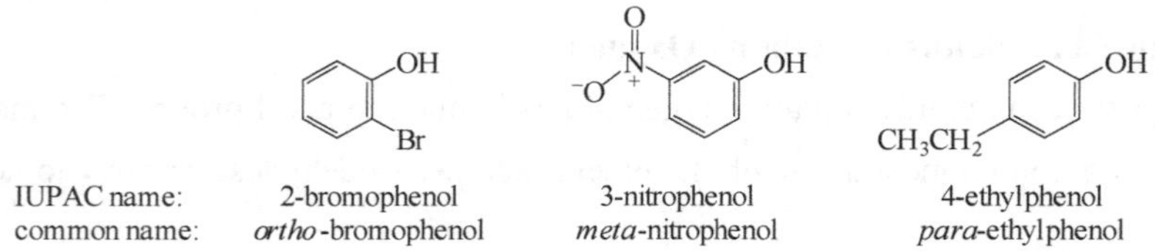

IUPAC name:	2-bromophenol	3-nitrophenol	4-ethylphenol
common name:	*ortho*-bromophenol	*meta*-nitrophenol	*para*-ethylphenol

Fig.1.32 Names of phenols

The methylphenols are called cresols, while the names of the benzenediols are based on their historical uses and sources rather than their structures. We generally use the systematic names of phenolic compounds (Fig.1.33).

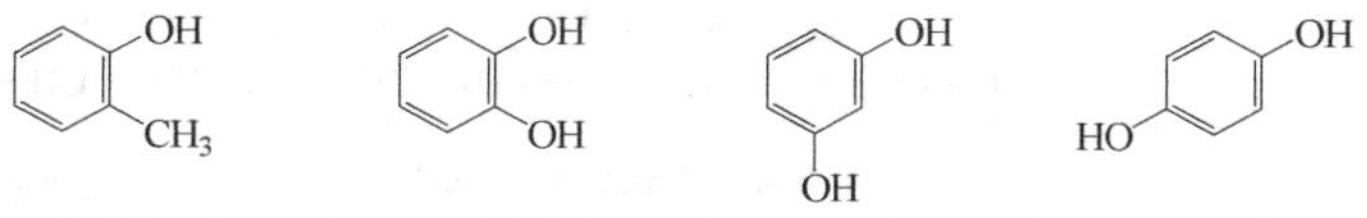

IUPAC name:	2- methylphenol	benzene-1,2-diol	benzene-1,3-diol	benzene-1,4-diol
common name:	*ortho*-cresol	catechol	resorcinol	hydroquinone

Fig.1.33 Names of benzenediols

(3) Ethers

Ethers are composed of two alkyl groups bonded to an oxygen atom. Like alcohols, ethers are much more polar than hydrocarbons, but ethers have no —OH groups, so they cannot form hydrogen bond with themselves. Ethers do form hydrogen bonds with hydrogen-bond donors such as alcohols, amines, and water, enhancing their solubility with these compounds. Ether names are often formed from the names of the alkyl groups and the word "ether". Diethyl ether is the common "ether" used for starting engines in cold weather and once used for surgical anesthesia (Fig.1.34).

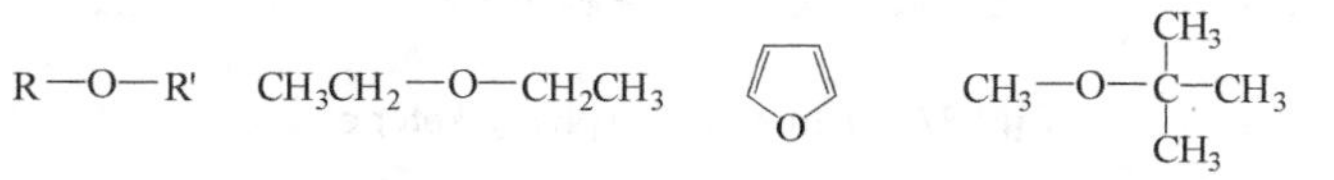

R—O—R' (ether); CH_3CH_2—O—CH_2CH_3 (diethyl ether); furan; CH_3—O—$C(CH_3)_2$—CH_3 (methyl *tert*-butyl ether)

Fig.1.34 Names of ethers

(4) Aldehydes and Ketones

The carbonyl group is the functional group for both aldehydes and ketones. A ketone has two alkyl groups bonded to the carbonyl group, an aldehyde has one alkyl group and a hydrogen atom bonded to the carbonyl group. Ketone names generally have the -one suffix, aldehyde names use either the -al suffix or the -aldehyde suffix (Fig.1.35).

The carbonyl group is strongly polar, and it can form hydrogen bonds with hydrogen-bond donors such as water, alcohols, and amines. Aldehydes and ketones containing up to four carbon atoms are miscible with water. Both acetone and acetaldehyde are miscible with water. Acetone, often used in nail polish remover, is a common solvent with low toxicity.

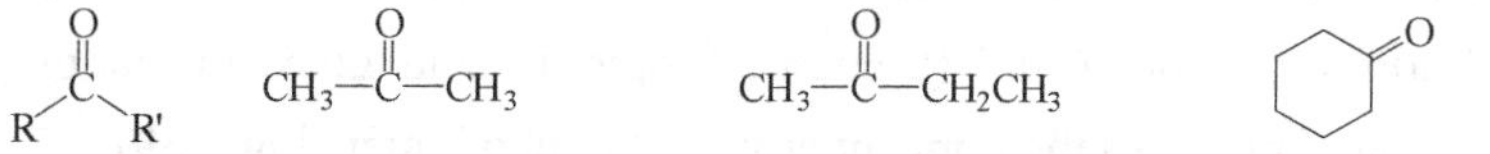

ketone; propan-2-one (acetone); butan-2- one (methyl ethyl ketone); cyclohexanone

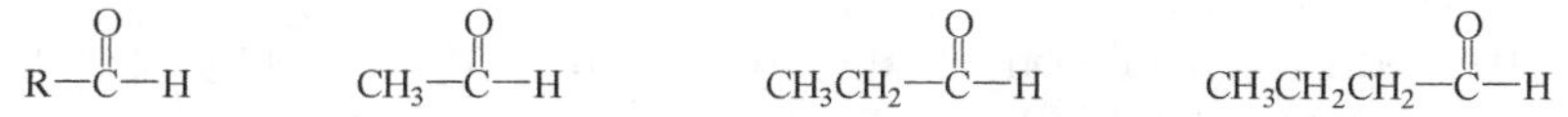

aldehyde; ethanal (acetaldehyde); propanal (propionaldehyde); butanal (butyraldehyde)

Fig.1.35 Names of aldehydes and ketones

Systematic names for aldehydes are derived by replacing the final -e of the alkane name with -al. An aldehyde carbon is at the end of a chain, so it is number 1. If the aldehyde group is a substituent of a large unit (usually a ring), the suffix -carbaldehyde is used (Fig.1.36).

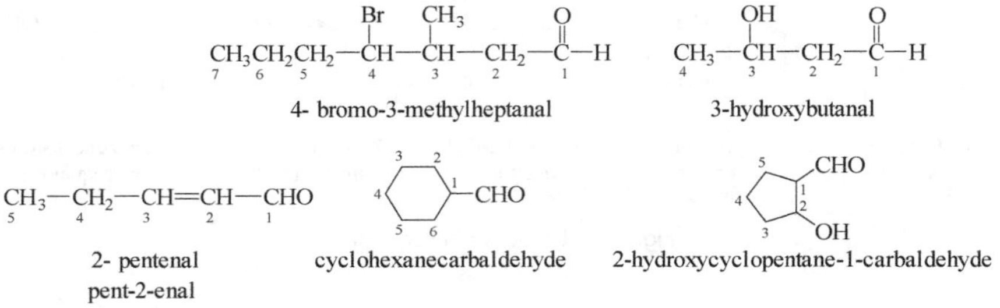

4- bromo-3-methylheptanal

3-hydroxybutanal

2- pentenal
pent-2-enal

cyclohexanecarbaldehyde

2-hydroxycyclopentane-1-carbaldehyde

Fig.1.36 Names of carbaldehydes

Some ketones have historical common names. Dimethyl ketone is always called acetone, and alkyl phenyl ketones are usually named as the acyl group followed by the suffix -phenone (Fig.1.37).

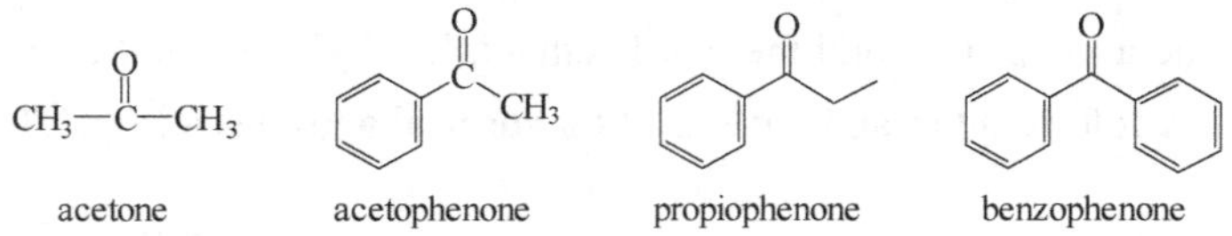

acetone　acetophenone　propiophenone　benzophenone

Fig.1.37 Names of alkyl phenyl ketones

(5) Carboxylic Acids

Carboxylic acids contain the carboxyl group, —COOH as their functional group. The general formula for carboxylic acid is RCO_2H. The carboxyl group combines a carbonyl group and a hydroxyl group, but this combination has different properties from those of ketones and alcohols. Carboxylic acids owe their acidity (pK_a of about 5) to the resonance-stabilized carboxylate anions formed by deprotonation. The following reaction shows the dissociation of a carboxylic acid(Fig.1.38).

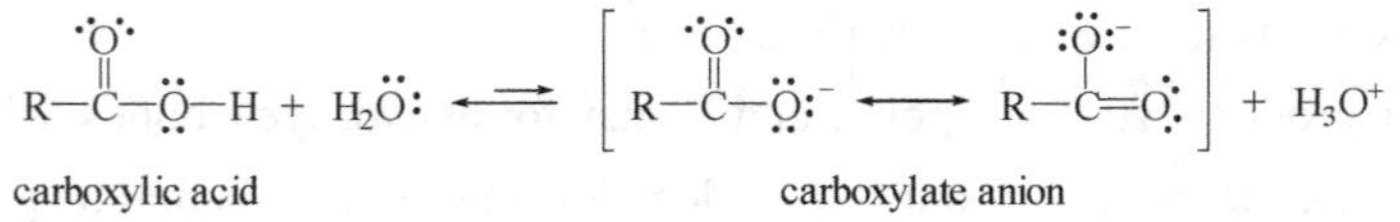

carboxylic acid　carboxylate anion

Fig.1.38 The resonance-stablized carboxylate anions formed by deprotonation

Formic acid was first isolated from ants, genus formica. Acetic acid, found in vinegar, gets its name from the Latin word for "sour" (acetum). Propionic acid gives the tangy flavor to sharp cheeses, and butyric acid provides the pungent aroma of rancid butter (Fig.1.39).

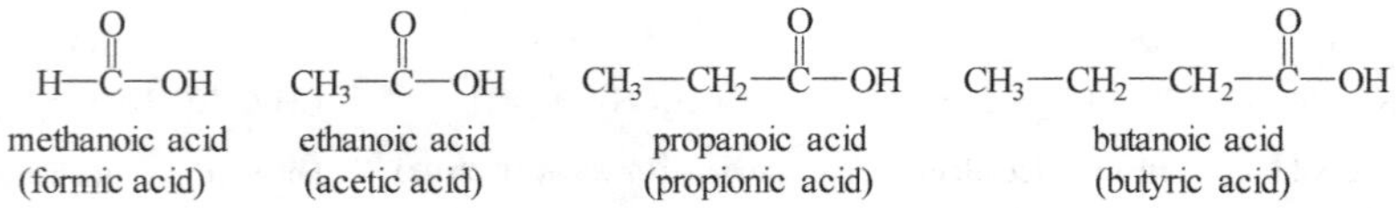

methanoic acid (formic acid)　ethanoic acid (acetic acid)　propanoic acid (propionic acid)　butanoic acid (butyric acid)

Fig.1.39 Names of acids

Carboxylic acids use the name of the alkane that corresponds to the longest continuous chain of carbon atoms. The final -e in the alkane name is replaced by the suffix -oic acid. The chain is numbered, starting with the carboxyl carbon atom, to give positions of substituents along the chain. In naming, the carboxyl group takes priority over any of the other functional groups we have discussed (Fig.1.40).

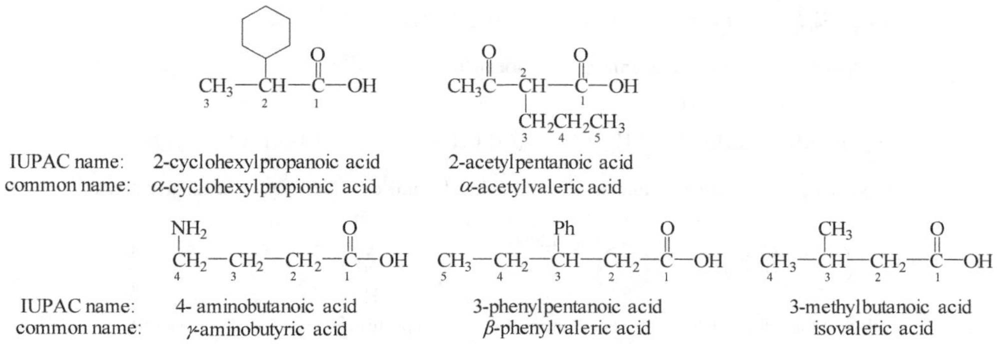

$$\overset{3}{CH_3}-\overset{2}{CH}(C_6H_{11})-\overset{1}{C}(=O)-OH \qquad CH_3C(=O)-\overset{2}{CH}(\overset{3}{C}H_2\overset{4}{C}H_2\overset{5}{C}H_3)-\overset{1}{C}(=O)-OH$$

IUPAC name:	2-cyclohexylpropanoic acid	2-acetylpentanoic acid
common name:	α-cyclohexylpropionic acid	α-acetylvaleric acid

$$\overset{4}{CH_2}(NH_2)-\overset{3}{CH_2}-\overset{2}{CH_2}-\overset{1}{C}(=O)-OH \qquad \overset{5}{CH_3}-\overset{4}{CH_2}-\overset{3}{CH}(Ph)-\overset{2}{CH_2}-\overset{1}{C}(=O)-OH \qquad \overset{4}{CH_3}-\overset{3}{CH}(CH_3)-\overset{2}{CH_2}-\overset{1}{C}(=O)-OH$$

IUPAC name:	4- aminobutanoic acid	3-phenylpentanoic acid	3-methylbutanoic acid
common name:	γ-aminobutyric acid	β-phenylvaleric acid	isovaleric acid

Fig.1.40 Names of carboxylic acids

(6) Carboxylic Acid Derivatives

Carboxylic acids are easily converted to a variety of acid derivatives. Each derivative contains the carbonyl group bonded to an oxygen or other electron-withdrawing element. Among these functional groups are acid chlorides, esters, and amides. All of these groups can be converted back to carboxylic acids by acidic or basic hydrolysis (Fig.1.41).

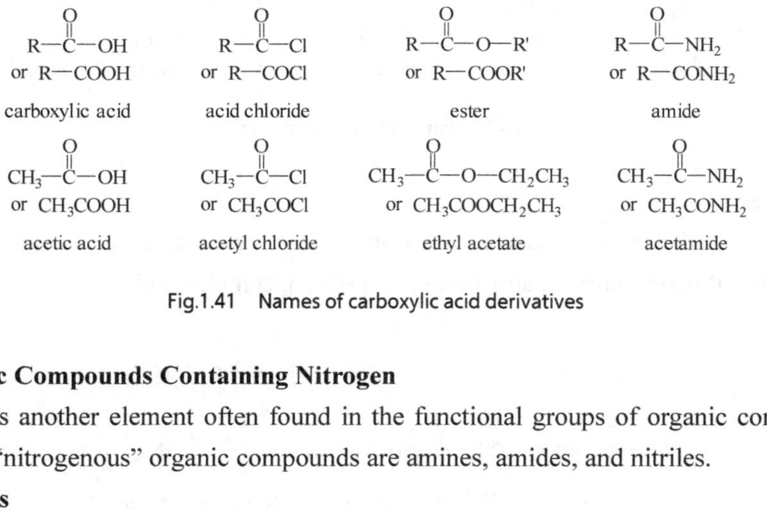

$R-C(=O)-OH$	$R-C(=O)-Cl$	$R-C(=O)-O-R'$	$R-C(=O)-NH_2$
or $R-COOH$	or $R-COCl$	or $R-COOR'$	or $R-CONH_2$
carboxylic acid	acid chloride	ester	amide
$CH_3-C(=O)-OH$	$CH_3-C(=O)-Cl$	$CH_3-C(=O)-O-CH_2CH_3$	$CH_3-C(=O)-NH_2$
or CH_3COOH	or CH_3COCl	or $CH_3COOCH_2CH_3$	or CH_3CONH_2
acetic acid	acetyl chloride	ethyl acetate	acetamide

Fig.1.41 Names of carboxylic acid derivatives

4. Organic Compounds Containing Nitrogen

Nitrogen is another element often found in the functional groups of organic compounds. The most common "nitrogenous" organic compounds are amines, amides, and nitriles.

(1) Amines

Amines are alkylated derivatives of ammonia. Like ammonia, amines are basic.

$$R-\ddot{N}H_2 + H_2O \rightleftharpoons R-\overset{+}{N}H_3 + OH^- \qquad K_b \approx 10^{-4}$$

Because of their basicity ("alkalinity"), naturally occurring amines are often called alkaloids. Simple amines are named by naming the alkyl groups bonded to nitrogen and adding the word "amine". The prefixes di-, tri- are used to describe two or three identical substituents. The structures of some simple amines are shown below, together with the structure of nicotine, a toxic alkaloid found in tobacco leaves (Fig.1.42).

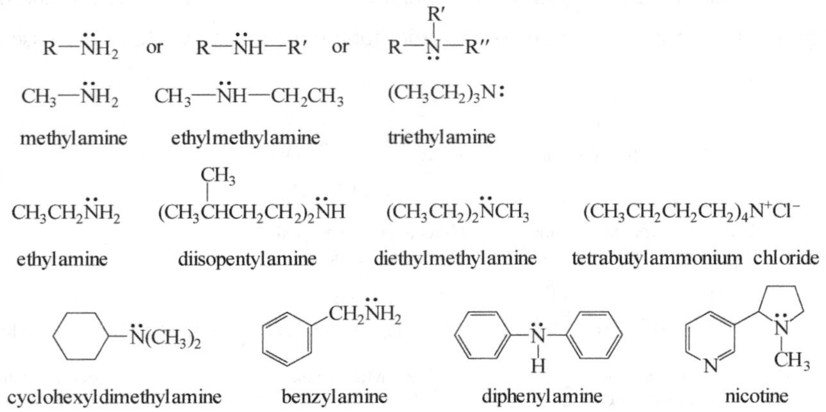

$R-\ddot{N}H_2$ or $R-\ddot{N}H-R'$ or $R-\ddot{N}(R')-R''$

$CH_3-\ddot{N}H_2$ methylamine

$CH_3-\ddot{N}H-CH_2CH_3$ ethylmethylamine

$(CH_3CH_2)_3N:$ triethylamine

$CH_3CH_2\ddot{N}H_2$ ethylamine

$(CH_3CH(CH_3)CH_2CH_2)_2\ddot{N}H$ diisopentylamine

$(CH_3CH_2)_2\ddot{N}CH_3$ diethylmethylamine

$(CH_3CH_2CH_2CH_2)_4N^+Cl^-$ tetrabutylammonium chloride

$\ddot{N}(CH_3)_2$ cyclohexyldimethylamine

$CH_2\ddot{N}H_2$ benzylamine

N H diphenylamine

N N CH_3 nicotine

Fig.1.42 Names of amines

Amines are classified as primary (1°), secondary (2°), or tertiary (3°), corresponding to one, two, or three alkyl or aryl groups bonded to nitrogen (Fig.1.43). In a heterocyclic amine, the nitrogen atom is part of an aliphatic or aromatic ring.

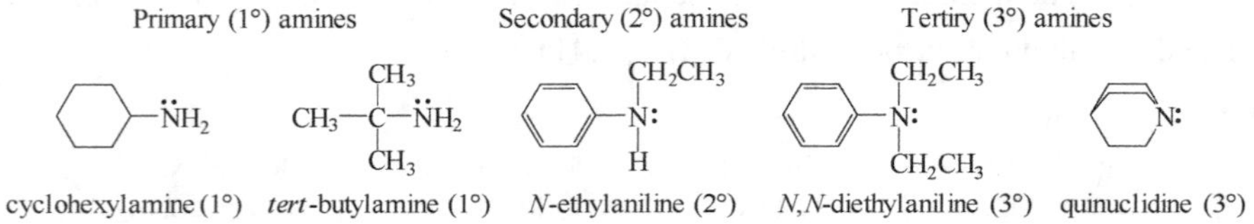

Primary (1°) amines | Secondary (2°) amines | Tertiry (3°) amines

$\ddot{N}H_2$ cyclohexylamine (1°)

$CH_3-C(CH_3)_2-\ddot{N}H_2$ *tert*-butylamine (1°)

CH_2CH_3 N: H *N*-ethylaniline (2°)

CH_2CH_3 N: CH_2CH_3 *N,N*-diethylaniline (3°)

N: quinuclidine (3°)

Fig.1.43 Names of heterocyclic amines

(2) Amides

Amides are acid derivatives that result from a combination of an acid with ammonia or an amine (Fig.1.44). Proteins have the structure of long-chain, complex amides.

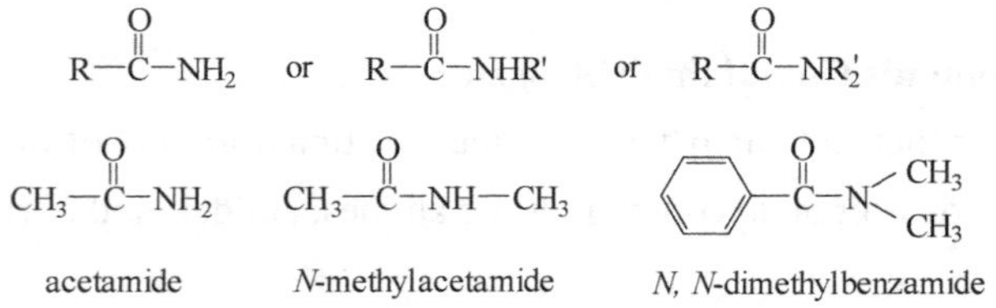

$R-C(=O)-NH_2$ or $R-C(=O)-NHR'$ or $R-C(=O)-NR'_2$

$CH_3-C(=O)-NH_2$ acetamide

$CH_3-C(=O)-NH-CH_3$ *N*-methylacetamide

$C(=O)-N(CH_3)_2$ *N, N*-dimethylbenzamide

Fig.1.44 Names of amides

Amides are among the most stable acid derivatives. The nitrogen atom of an amide is not as basic as the nitrogen of an amine because of the electron-withdrawing effect of the carbonyl group. The following resonance forms help to show why amides are very weak bases (Fig.1.45).

$$\left[R-\overset{\overset{\large \cdot\ddot{O}\cdot}{\|}}{C}-\ddot{N}H_2 \longleftrightarrow R-\overset{\overset{:\ddot{O}:^-}{|}}{C}=\overset{+}{N}H_2 \right]$$

very weak base

Fig.1.45 Resonance forms of amides

Amides form particularly strong hydrogen bonds, giving them high melting points and high boiling points. The strongly polarized amide hydrogen N—H forms unusually strong hydrogen bonds with the carbonyl oxygen that carries a partial negative charge in the polarized resonance form shown above.

(3) Nitriles

A nitrile is a compound containing the cyano group. The cyano group is an example of sp hybridized bonding (Fig.1.46). The cyano group is strongly polar by virtue of the triple bond, and most small nitriles are somewhat soluble in water. Acetonitrile is miscible with water.

$R-C\equiv N:$	$CH_3-C\equiv N:$	$CH_3CH_2-C\equiv N:$	$C_6H_5-C\equiv N:$
nitrile	acetonitrile	propionitrile	benzonitrile

Fig.1.46 Names of nitriles

5. Nomenclature of Chiral Organic Compounds

What is the difference between your left hand and your right hand? They look similar, yet a left-hand glove does not fit the right hand. The same principle applies to your feet. They look almost identical, yet the left shoe fits painfully on the right foot. The relationship between your two hands or your two feet is that they are nonsuperimposable (nonidentical) mirror images of each other. Objects that have left-hand and right-hand forms are called chiral, the Greek word for "hand". The most common feature (but not the only one) that lends chirality is a carbon atom that is bonded to four different groups. Such a carbon atom is called an asymmetric carbon atom or a chiral carbon atom, and is often designated by an asterisk *. An asymmetric carbon atom is the most common example of a chirality center (or chiral center), the IUPAC term for any atom holding a set of ligands in a spatial arrangement that is not superimposable on its mirror image. Chirality centers belong to an even broader group called stereocenters. A stereocenter (or stereogenic atom) is any atom at which the interchange of two groups gives a stereoisomer.

Any asymmetric carbon has two possible (mirror-image) spatial arrangements, which we call configurations. The alanine enantiomers represent the two possible arrangements of its four groups around the asymmetric carbon atom. If we can name the two configurations of any asymmetric carbon atom, then we have a way of specifying and naming the enantiomers of alanine or any other chiral compound (Fig.1.47).

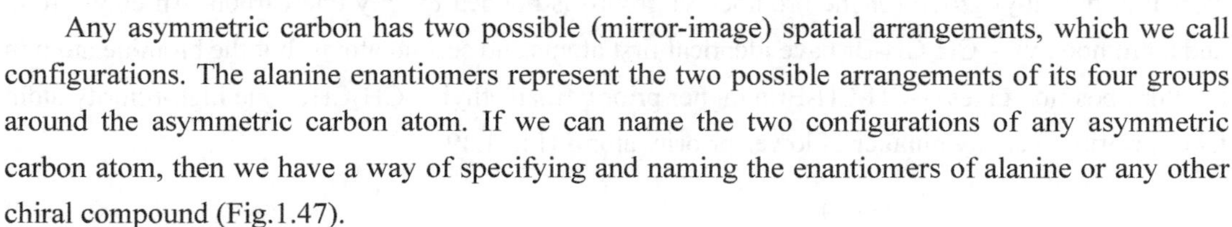

The Cahn-Ingold-Prelog convention is the most widely accepted system for naming the configurations of chirality centers. Each asymmetric carbon atom is assigned a letter (*R*) or (*S*) based on its three-dimensional configuration. To determine the name, we follow a two-step procedure that assigns "priorities" to the four substituents and then assigns the name based on the relative positions of these substituents. Here is the procedure.

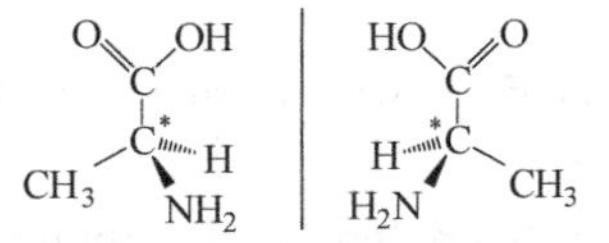

natural alanine　　unnatural alanine

Fig.1.47　Names of the enantiomers of alanine

Step 1: Assign a relative "priority" to each group bonded to the asymmetric carbon. We speak of group 1 as having the highest priority, group 2 second, group 3 third, and group 4 as having the lowest priority (Fig.1.48).

Look at the first atom of the group—the atom bonded to the asymmetric carbon. Atoms with higher atomic numbers receive higher priorities. For example, if the four groups bonded to an asymmetric carbon atom were H, CH_3, NH_2 and F, the fluorine atom (atomic number 9) would have the highest priority, followed by the nitrogen atom of the NH_2 group (atomic number 7) and the carbon atom of the methyl group (atomic number 6). Note that we look only at the atomic number of the atom directly attached to the asymmetric carbon, not the entire group. Hydrogen would have the lowest priority. With different isotopes of the same element, the heavier isotopes have higher priorities. For example, tritium (^3H) receives a higher priority than deuterium (^2H) followed by hydrogen (^1H). Examples of priority for atoms bonded to an asymmetric carbon: I > Br > Cl > S > F > O > N > ^{13}C > ^{12}C > Li > ^3H > ^2H > ^1H.

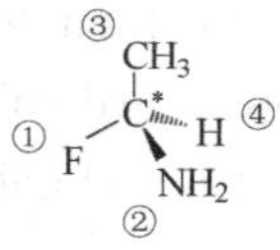

Fig.1.48　The configurations of chiral carbon

In case of ties, use the next atoms along the chain of each group as tiebreakers. For example, we assign a higher priority to isopropyl —$CH(CH_3)_2$ than to ethyl —CH_2CH_3 or bromoethyl —CH_2CH_2Br. The first carbon in the isopropyl group is bonded to two carbons, while the first carbon in the ethyl group (or the bromoethyl group) is bonded to only one carbon. An ethyl group and a bromoethyl —CH_2CH_2Br have identical first atoms and second atoms, but the bromine atom in the third position gives —CH_2CH_2Br a higher priority than ethyl —CH_2CH_3. One high-priority atom takes priority over any number of lower-priority atoms (Fig.1.49).

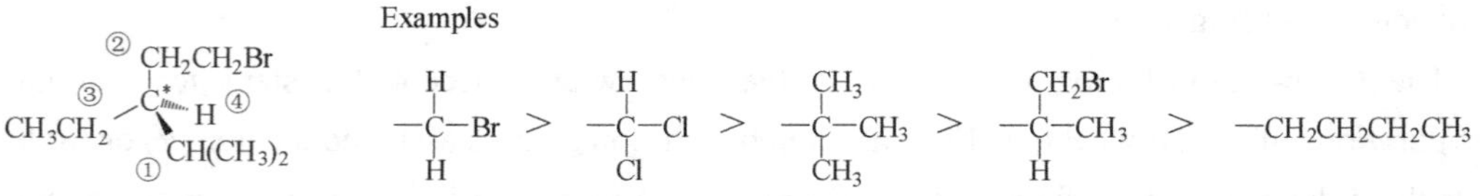

Examples

$-CH_2Br$ > $-CHCl_2$ > $-C(CH_3)_3$ > $-CH(CH_3)CH_2Br$ > $-CH_2CH_2CH_2CH_3$

Fig.1.49　The order of decreasing priority in the asymmetric carbon

Treat double and triple bonds as if each were a bond to a separate atom. For this method, imagine that each π bond is broken and the atoms at both ends duplicated. Note that when you break a bond, you always add two imaginary atoms. (Imaginary atoms are circled in Fig.1.50)

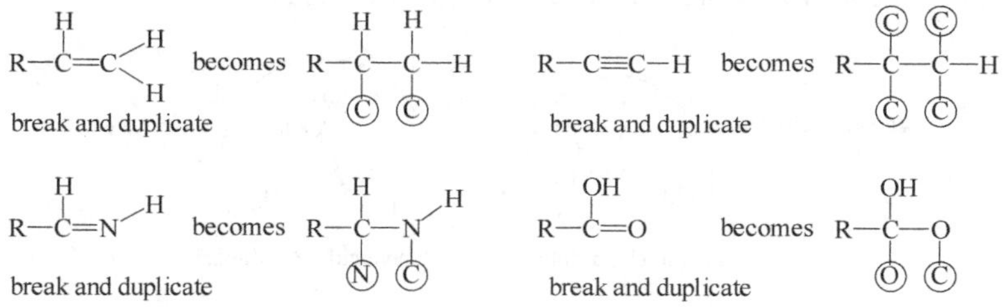

R—CH=CH₂ becomes R—CH(Ⓒ)—CH₂(Ⓒ) (break and duplicate)

R—C≡C—H becomes R—C(Ⓒ)(Ⓒ)—C(Ⓒ)(Ⓒ)—H (break and duplicate)

R—CH=N—H becomes R—CH(Ⓝ)—N(Ⓒ)—H (break and duplicate)

R—C(OH)=O becomes R—C(OH)(Ⓞ)—O(Ⓒ) (break and duplicate)

Fig.1.50 The order of decreasing priority in this chiral carbon

Step 2: Using a three-dimensional drawing or a model, put the fourth-priority group away from you and view the molecule with the first, second, and third priority groups radiating toward you like the spokes of a steering wheel. Draw an arrow from the first-priority group, through the second, to the third. If the arrow points clockwise, the asymmetric carbon atom is called (*R*) (Latin, rectus, "upright"). If the arrow points counter-clockwise, the chiral carbon atom is called (*S*) (Latin, sinister, "left"). Alternatively, you can draw the arrow and imagine turning a car's steering wheel in that direction. If the car would go to the left, the asymmetric carbon atom is designated (*S*). If the car would go to the right, the asymmetric carbon atom is designated (*R*) (Fig.1.51).

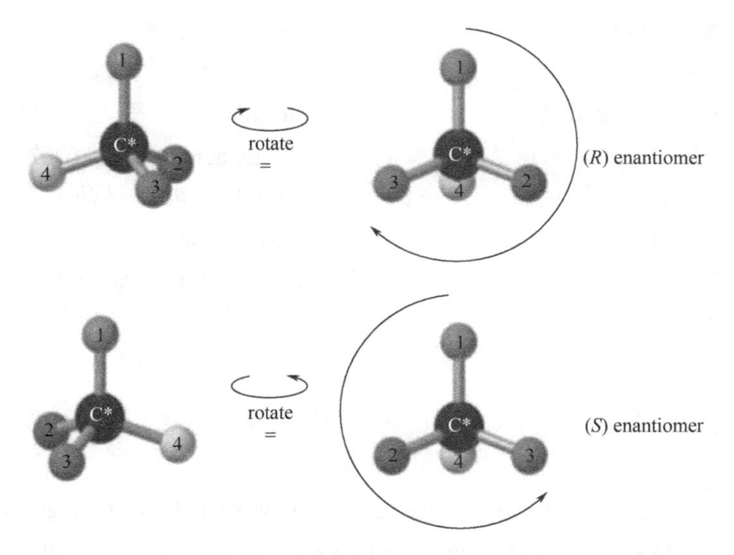

Fig.1.51 The three-dimensional drawing model of stereoisomers

Let us use the enantiomers of alanine as an example. The naturally occurring enantiomer is the one on the left, determined to have the (*S*) configuration. Of the four atoms attached to the asymmetric carbon in alanine, nitrogen has the largest atomic number, giving it the highest priority. Next is the —COOH carbon atom, since it is bonded to oxygen atoms. The third is the methyl group, followed by the hydrogen atom. When we position the natural enantiomer with its hydrogen atom pointing away from us, the arrow from $—NH_2$ to —COOH to $—CH_3$ points counter-clockwise. Thus, the naturally occurring enantiomer of alanine has the (*S*) configuration. Make models of these enantiomers to illustrate how they are named (*R*) and (*S*) (Fig.1.52).

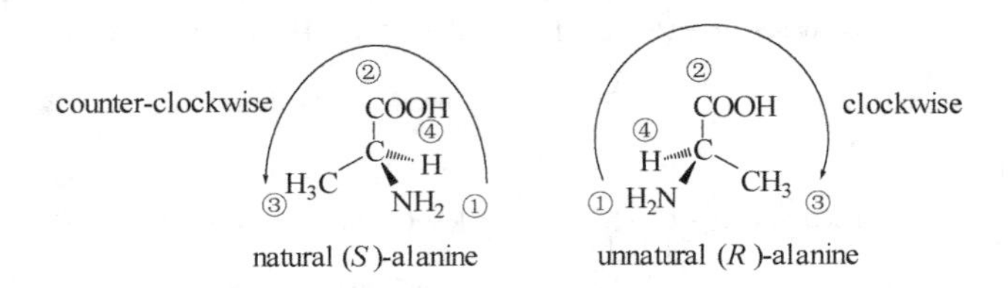

Fig.1.52 The configuration of alanine enantiomers

Words and Expressions

hydrocarbon *n.* 烃；碳氢化合物
cycloalkane *n.* 环烷烃
isomer *n.* 同分异构体；同质异能素
isopropyl *n.* 异丙基
isopentane *n.* 异戊烷
neopentane *n.* 新戊烷
alkene *n.* 烯烃；链烯
alkyne *n.* 炔烃
acetylene *n.* 乙炔；电石气
benzene *n.* 苯
toluene *n.* 甲苯
xylene *n.* 二甲苯
amine *n.* 胺
amide *n.* 酰胺
nitrile *n.* 腈
aldehyde *n.* 醛
ketone *n.* 酮
alcohol *n.* 醇
phenol *n.* 苯酚；酚
ether *n.* 醚
acyl *n.* 酰基
carboxyl *n.* 羧基
formic acid 甲酸；蚁酸
acetic acid 乙酸；醋酸
chiral *adj.* 手性的
enantiomer *n.* 对映体；对映异构体
asymmetric *adj.* 不对称的；非对称的

Notes

➢ Just as a generic alkyl group substituent is represented by R, a generic aryl group is represented by Ar. When a benzene ring serves as a substituent, it is called a phenyl group, abbreviated Ph.

参考译文：就像一般的烷基取代基用 R 表示一样，芳基用 Ar 表示。当一个苯环作为取代基时，称为苯基，缩写为 Ph。

➢ Alkanes undergo few reactions because they have no functional group, the part of the molecule where reactions usually occur.

参考译文：通常官能团是分子中能发生反应的部分，但由于烷烃分子本身没有官能团，因此它们很难进行反应。

➢ The rigidity and lack of rotation of carbon-carbon double bonds give rise to *cis-trans* isomerism, also called geometric isomerism.

参考译文：碳碳双键的刚性和无法旋转导致了顺反异构，也称为几何异构。

➢ Alcohols are among the most polar organic compounds because the hydroxyl group is strongly polar and can participate in hydrogen bonding. Some of the simple alcohols like ethanol and methanol are miscible (soluble in all proportions) with water.

参考译文：由于羟基具有很强的极性，并且可以参与氢键作用，因此醇类是极性最大的一类有机化合物。一些简单的醇，如乙醇和甲醇，可与水混溶（任意比互溶）。

➢ Amines are classified as primary (1°), secondary (2°), or tertiary (3°), corresponding to one, two, or three alkyl or aryl groups bonded to nitrogen. In a heterocyclic amine, the nitrogen atom is part of an aliphatic or aromatic ring.

参考译文：胺类化合物可分为伯胺（1°）、仲胺（2°）或叔胺（3°），其对应于一个、两个或三个与氮相连的烷基或芳基。在杂环胺类化合物中，氮原子是脂肪环或芳香环的一部分。

➢ Amides are acid derivatives that result from a combination of an acid with ammonia or an amine. Proteins have the structure of long-chain, complex amides.

参考译文：酰胺是酸与氨或胺结合产生的酸衍生物。蛋白质具有长链和复杂酰胺的结构。

➢ Chirality centers belong to an even broader group called stereocenters. A stereocenter (or stereogenic atom) is any atom at which the interchange of two groups gives a stereoisomer.

参考译文：手性中心属于立体中心（更广泛的一类）。立体中心（或立体原子）是指两个基团交换形成立体异构体的任何原子。

Exercises

1. Denominate the following compounds

(1) 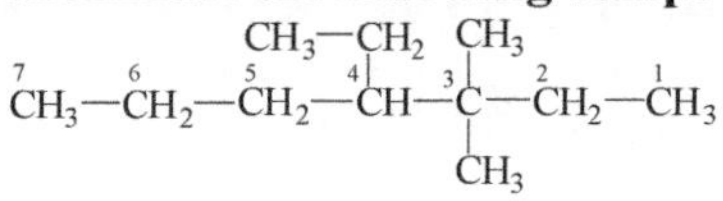 $\overset{7}{C}H_3-\overset{6}{C}H_2-\overset{5}{C}H_2-\overset{4}{C}H(CH_2-CH_3)-\overset{3}{C}(CH_3)_2-\overset{2}{C}H_2-\overset{1}{C}H_3$

(2) $\overset{4}{C}H_3-\overset{3}{C}H(CH_3)-\overset{2}{C}H=\overset{1}{C}H_2$

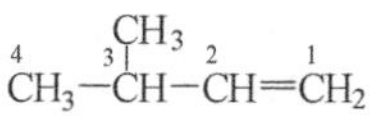

(3) 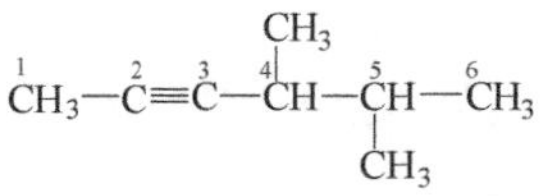 $\overset{1}{C}H_3-\overset{2}{C}\equiv\overset{3}{C}-\overset{4}{C}H(CH_3)-\overset{5}{C}H(CH_3)-\overset{6}{C}H_3$

(4) $C_6H_5-CO-CH_3$ (benzene ring–CO–CH₃)

(5) cyclohexane ring bearing COOH and OH (1,3-positions)

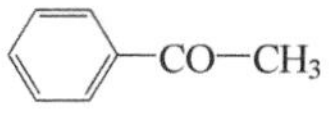

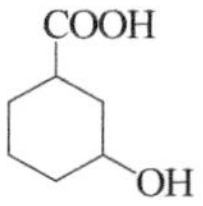

(6) $CH_3-O-CH_2-CH_3$

2. Draw structural formulas

(1) 1-bromo-2-chloro-4-ethylbenzene

(2) 4-bromo-1,2-dimethylbenzene

(3) 2,4,6-trinitrotoluene

(4) 4-phenyl-1-pentene

(5) *p*-cresol

(6) 2,4-dichlorophenol

3. Translate the following into Chinese

(1) We might try to name this compound as "cyclohexylethane", but we should treat the larger fragment as the parent compound (cyclohexane), and the smaller group as the alkyl group (ethyl).

(2) Many organic compounds contain oxygen atoms bonded to alkyl groups. The major classes of oxygen-containing compounds are alcohols, ethers, ketones, aldehydes, carboxylic acids, and acid derivatives.

(3) A nitrile is a compound containing the cyano group. The cyano group is an example of sp hybridized bonding. The cyano group is strongly polar by virtue of the triple bond, and most small nitriles are somewhat soluble in water. Acetonitrile is miscible with water.

(4) If two similar groups bonded to the carbons of the double bond are on the same side of the bond, the alkene is *cis* isomer. If the similar groups are on opposite sides of the bond, the alkene is *trans* isomer.

(5) Some ordinary aromatic hydrocarbons, such as benzene, toluene, xylene and naphthalene, are mainly obtained from coal and petroleum.

(6) The carbonyl group is the functional group for both aldehydes and ketones. A ketone has two alkyl groups bonded to the carbonyl group, an aldehyde has one alkyl group and a hydrogen atom bonded to the carbonyl group. Ketone names generally have the -one suffix, aldehyde names use either the -al suffix or the -aldehyde suffix.

4. Discuss the following questions

We say that naphthalene, anthracene, phenanthrene, and benzo[*a*]pyrene are polynuclear aromatic hydrocarbons. In this context, what does "polynuclear" mean? What does "aromatic" mean? What does "hydrocarbon" mean?

PART 2

Structure Characterization of Organic Compounds

UNIT 1 Ultraviolet-visible Spectroscopy

1. Colour

As well known, Newton's discovery that the spectral colours are the components of the "colour" white is one of the great discoveries of physics. He has no doubts of being able to obtain pure white from a mixture of four or five spectral rays. Then, he presents his colour circle (Fig.2.1). The spectrum is curved into a circle, so that red lies next to violet, and white is placed in the middle. In the spectrum, seven colours are distinguished which gradually merge into each other. With the development of science, ultraviolet-visible spectroscopy is being introduced to well define these "colours".

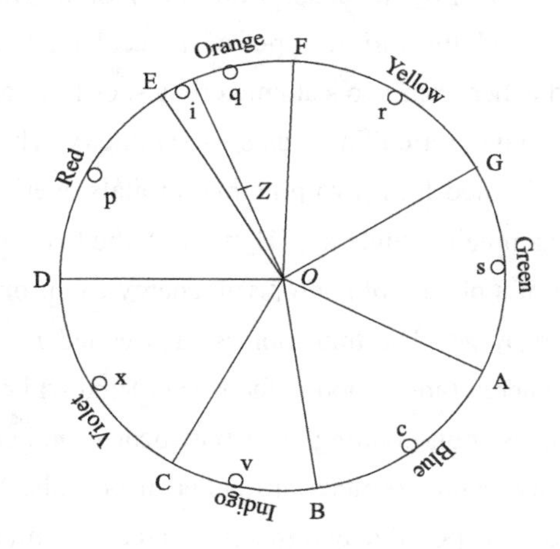

Fig.2.1 Newton's colour circle (1704)

(The spectrum has been bent so that violet and red meet. White is at the center.)

2. Ultraviolet-visible Spectroscopy

Ultraviolet-visible spectroscopy or ultraviolet-visible spectrophotometry (UV-Vis or UV/Vis) refers to absorption spectroscopy or reflectance spectroscopy in part of the ultraviolet and the full adjacent visible spectral regions. This means it uses light in the visible and adjacent ranges. The absorption or reflectance in the visible range directly affects the perceived colour of the chemicals involved. In this region of the electromagnetic spectrum, atoms and molecules undergo electronic transitions. Absorption spectroscopy is complementary to fluorescence spectroscopy, in that fluorescence deals with transitions from the excited state to the ground state, while absorption measures transitions from the ground state to the excited state (see Fig.2.2).

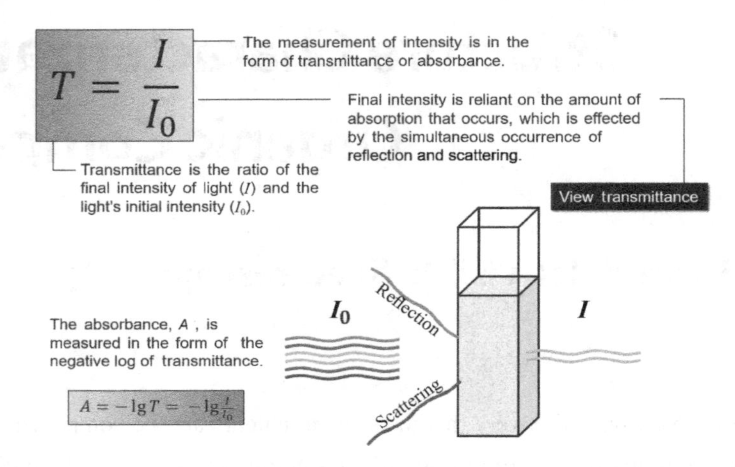

Fig.2.2 Absorption of light

3. Types of Electronic Transitions in Polyatomic Molecules

An electronic transition consists of promoting an electron from an orbital of a molecule in the ground state to an unoccupied orbital by the absorption of a photon. The molecule is then said to be in an excited state. Let us first recall the various types of molecular orbitals.

A σ orbital can be formed either from two s atomic orbitals, or from one s and one p atomic orbital, or from two p atomic orbitals having a collinear axis of symmetry. The bond formed in this way is called a σ bond. A π orbital is formed from two p atomic orbitals overlapping laterally. The resulting bond is called a π bond. For example in ethylene ($CH_2{=}CH_2$), the two carbon atoms are linked by one σ and one π bond. Absorption of a photon of appropriate energy can promote one of the π electrons to an antibonding orbital denoted by π*. The transition is then called π → π*. The promotion of a σ electron requires much higher energy (absorption in the far UV) and will not be considered here.

A molecule may also possess non-bonding electrons located on heteroatoms such as oxygen or nitrogen. The corresponding molecular orbitals are called n orbitals. Promotion of a non-bonding electron to an antibonding orbital is possible and the associated transition is denoted by n → π*. The energy of these electronic transitions is generally in the following order

$$n \rightarrow \pi^* < \pi \rightarrow \pi^* < n \rightarrow \sigma^* < \sigma \rightarrow \pi^* < \sigma \rightarrow \sigma^*$$

To illustrate these energy levels, Fig.2.3 shows formaldehyde as an example, with all the possible transitions. The n → π* transition deserves further attention. Upon excitation, an electron is removed from the oxygen atom and goes into the π* orbital localized half on the carbon atom and half on the oxygen atom. The n → π* excited state thus has a charge transfer character, as shown by an increase in the dipole moment of about 2D with respect to the ground state dipole moment of C=O (3D).

In absorption and fluorescence spectroscopy, two important types of orbitals are considered: the highest occupied molecular orbitals (HOMO) and the lowest unoccupied molecular orbitals (LUMO). Both of these refer to the ground state of the molecule. For instance, in formaldehyde, the HOMO is the n orbital and the LUMO is the π^* orbital (see Fig.2.3).

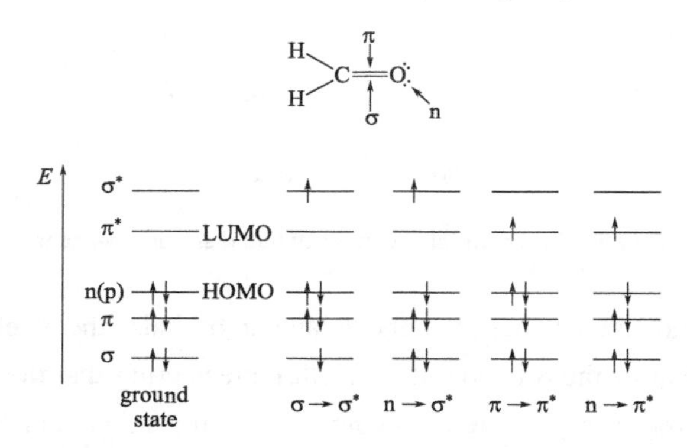

Fig.2.3 Energy levels of molecular orbitals in formaldehyde (HOMO, LUMO) and possible electronic transitions

When one of the two electrons of opposite spins (belonging to a molecular orbital of a molecule in the ground state) is promoted to a molecular orbital of higher energy, its spin is in principle unchanged so that the total spin quantum number ($S = \sum S_i = +\frac{1}{2}$ or $-\frac{1}{2}$) remains equal to zero. Because the multiplicities of both the ground and excited states ($M = 2S + 1$) is equal to 1, both are called singlet state (usually denoted S_0 for the ground state, and S_1, S_2... for the excited states) (Fig.2.4). The corresponding transition is called a singlet-singlet transition. It will be shown later that a molecule in a singlet excited state may undergo conversion into a state where the promoted electron has changed its spin, because there are two electrons with parallel spins, the total spin quantum number is 1 and the multiplicity is 3. Such a state is called a triplet state because it corresponds to three states of equal energy. According to Hund's rule, the triplet state has lower energy than that of the singlet state of the same configuration.

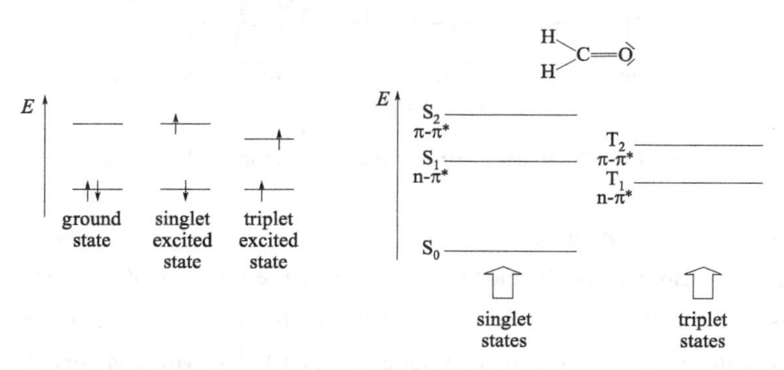

Fig.2.4 Distinction between singlet and triplet states, using formaldehyde as an example

In a molecule such as formaldehyde, the bonding and non-bonding orbitals are localized (like the bonds) between pairs of atoms. Such a picture of localized orbitals is valid for the σ orbitals of single bonds and for the π orbitals of isolated double bonds, but it is no longer adequate in the case of alternate single and double carbon-carbon bonds (in so-called conjugated systems). In fact, the overlap of the π orbitals allows the electrons to be delocalized over the whole system (resonance effect). Butadiene and benzene are the simplest cases of linear and cyclic conjugated systems, respectively (Fig.2.5).

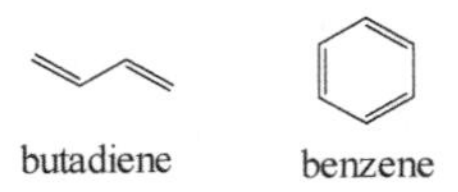

Fig.2.5 The chemical structures of butadiene and benzene

Because there is no overlap between the σ and π orbitals, the π electron system can be considered as independent of the σ bonds. It is worth remembering that the greater extent of the π electron system, the lower energy of the low-lying π → π* transition, and consequently, the larger wavelength of the corresponding absorption band (Fig.2.6). This rule applies to linear conjugated systems (polyenes) and cyclic conjugated systems (aromatic molecules).

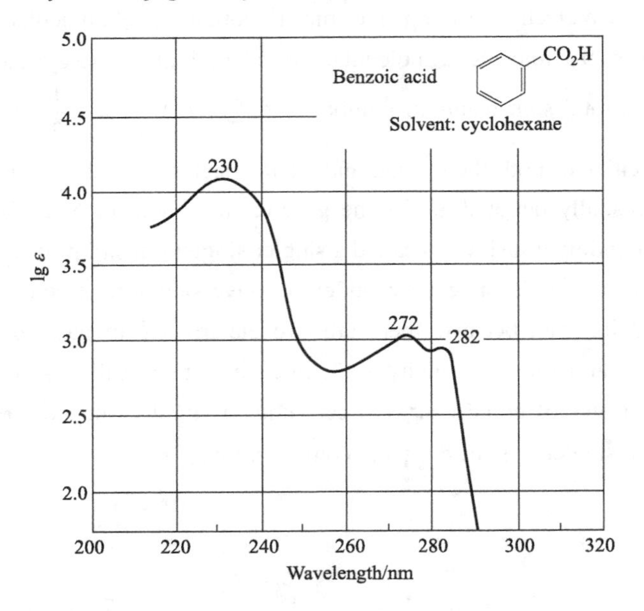

Fig.2.6 The actual spectrum which corresponds to these data

4. The Franck-Condon Principle

According to the Born-Oppenheimer approximation, the motions of electrons are much more rapid than those of the nuclei (i.e. the molecular vibrations). Promotion of an electron to an antibonding molecular orbital upon excitation takes about 10^{-15} s, which is very quick compared to the characteristic time for molecular vibrations (10^{-12}~10^{-10} s). This observation is the basis of the Franck-Condon principle: an electronic transition is most likely to occur without changes in the positions of the nuclei in the molecular entity and its environment. The resulting state is called a

Franck-Condon state, and the transition is called vertical transition, as illustrated by the energy diagram of Fig.2.7 in which the potential energy curve as a function of the nuclear configuration (internuclear distance in the case of a diatomic molecule) is represented by a Morse function.

At room temperature, most of the molecules are in the lowest vibrational level of the ground state (according to the Boltzmann distribution). In addition to the "pure" electronic transition called the 0-0 transition, there are several vibronic transitions whose intensities depend on the relative position and shape of the potential energy curves (Fig.2.7).

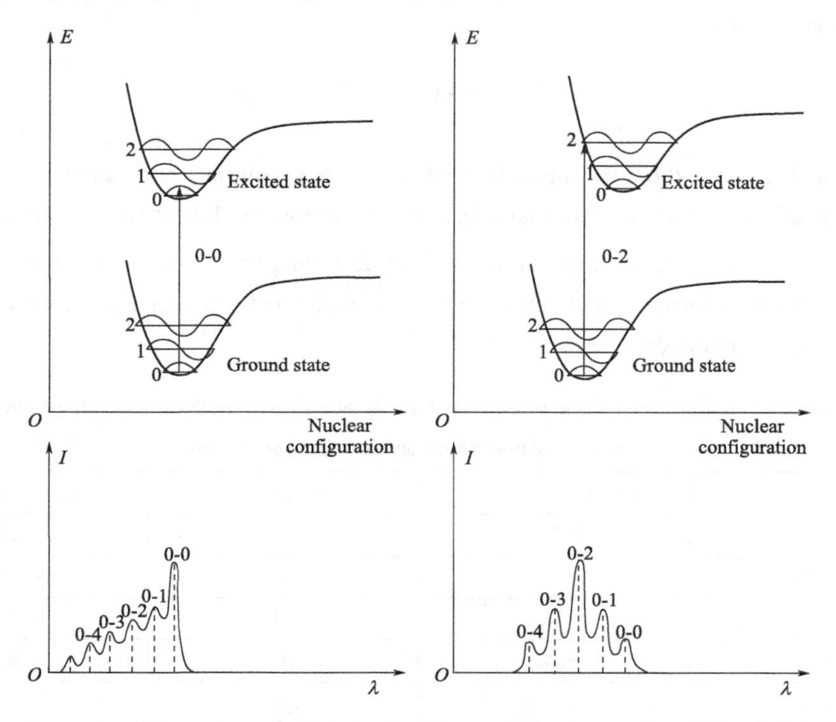

Fig.2.7 Top: Potential energy diagrams with vertical transitions (Franck-Condon principle); Bottom: Shape of the absorption bands

Note: The vertical broken lines represent the absorption lines that are observed for a vapor, whereas broadening of the spectra is expected in solution (solid line).

The width of a band in the absorption spectrum of a chromophore located in a particular microenvironment is a result of two effects: homogeneous and inhomogeneous broadening. Homogeneous broadening is due to the existence of a continuous set of vibrational sublevels in each electronic state. Inhomogeneous broadening results from the fluctuations of the structure of the solvation shell surrounding the chromophore. Such broadening effects also exist for emission bands in fluorescence spectra.

Due to the effect of substitution or a change in environment, shifts occur in absorption spectra. Note that a shift to longer wavelengths is called a bathochromic shift (informally referred to as a red-shift). A shift to shorter wavelengths is called a hypsochromic shift (informally referred to as a blue-shift). An increase in the molar absorption coefficient is called the hyperchromic effect, whereas the opposite is the hypochromic effect.

5. The Beer-Lambert Law

The Beer-Lambert law states that the absorbance of a solution is directly proportional to the concentration of the absorbing species in the solution and the path length. Thus, for a fixed path length, UV/Vis spectroscopy can be used to determine the concentration of the absorber in a solution. It is necessary to know how quickly the absorbance changes with concentration. Figures can be taken from references (tables of molar extinction coefficients), or more accurately, determined from a calibration curve

$$A(\lambda) = a_0 + \frac{I_\lambda^0}{I_\lambda} = \varepsilon(\lambda) lc \tag{2.1}$$

Where I_λ^0 and I_λ are the light intensities of the beams entering and leaving the absorbing medium, respectively. $\varepsilon(\lambda)$ is the molar (decadic) absorption coefficient (commonly expressed in $L\cdot mol^{-1}\cdot cm^{-1}$), c is the concentration (in $mol\cdot L^{-1}$) of absorbing species and l is the absorption path length (thickness of the absorbing medium) (in cm). Table 2.1 lists the molar absorption coefficients of some common compounds.

Table 2.1 Examples of molar absorption coefficients (at the wavelength corresponding to the maximum of the absorption band of lower energy)

Compound	$\varepsilon(\lambda)/(L\cdot mol^{-1}\cdot cm^{-1})$	Compound	$\varepsilon(\lambda)/(L\cdot mol^{-1}\cdot cm^{-1})$
Benzene	≈200	Acridine	≈12000
Phenol	≈2000	Biphenyl	≈16000
Carbazole	≈4200	Bianthryl	≈24000
1-Naphthol	≈5400	Acridine orange	≈30000
Indole	≈5500	Perylene	≈34000
Fluorene	≈9000	Eosin Y	≈90000
Anthracene	≈10000	Rhodamine B	≈105000
Quinine sulfate	≈10000		

The Beer-Lambert law is useful for characterizing many compounds. For example, a UV/Vis spectrophotometer may be used as a detector for HPLC. The presence of an analyte gives a response assumed to be proportional to concentration. For accurate results, the instrument's response to the analyte in the unknown should be compared with the response to a standard. This is very similar to the use of calibration curves. The response (e.g., peak height) for a particular concentration is known as the response factor.

6. Instrumentation

The typical UV/Vis spectrophotometer consists of a light source, a monochromator, and a detector. The light source is usually a deuterium lamp, which emits electromagnetic radiation in the ultraviolet region of the spectrum. A second light source, a tungsten lamp, is used for wavelengths in the visible region of the spectrum. The monochromator is a diffraction grating, to spread the beam of light into its component wavelengths. A system of slits focuses the desired wavelength on the sample cell. The light which passes through the sample cell reaches the detector, which records the intensity of the transmitted light (I). The detector is generally a photo-multiplier tube, although in modern instruments photodiodes are also used. In a typical double-beam instrument, the light emanating from the light source is split into two beams, the sample beam and the reference beam. When there is no sample cell in the reference beam, the detected light is taken to be equal to the intensity of light entering the sample (I_0).

The sample cell must be constructed of a material which is transparent to the electromagnetic radiation being used in the experiment. For spectra in the visible range of the spectrum, cells composed of glass or plastic are generally suitable. For measurements in the ultraviolet region of the spectrum, however, glass and plastic cannot be used because they absorb ultraviolet radiation. Instead, cells made of quartz must be used, since quartz does not absorb radiation in this region.

The instrument design just described is quite suitable for measurement at only one wavelength. If a complete spectrum is desired, this type of instrument has some deficiencies. A mechanical system is required to rotate the monochromator and provide a scan of all desired wavelengths. This type of system operates slowly, and therefore considerable time is required to record a spectrum. A modern improvement on the traditional spectrophotometer is the diode-array spectrophotometer. This type of detector has no moving parts and can record spectra very quickly. Furthermore, its output can be passed to a computer, which can process the information and provide a variety of useful output formats. Since the number of photodiodes is limited, the speed and convenience described here are obtained at some small cost in resolution.

Words and Expressions

fluorescence *n.* 荧光
unoccupied orbital 空轨道
axis of symmetry 对称轴
antibonding orbital 反键轨道
heteroatom *n.* 杂原子
formaldehyde *n.* 甲醛
spin quantum number 自旋量子数
multiplicity *n.* 多样性；多重性
vibronic *adj.* 电子振动的
potential energy curve 势能曲线
chromophore *n.* 发色团
spectrophotometer *n.* 分光光度计
electromagnetic radiation 电磁发射
homogeneous *adj.* 同质的；均相的
hypsochromic shift 蓝移
bathochromic shift 红移

Notes

➢ It is worth remembering that the greater the extent of the π electron system, the lower energy of the low-lying $\pi \rightarrow \pi^*$ transition, and consequently, larger wavelength of the corresponding absorption band.

参考译文：值得注意的是，π电子的体系越大，π→π*跃迁的能量越低，其相应的吸收带波长也就越长。

➢ An electronic transition is most likely to occur without changes in the positions of the nuclei in the molecular entity and its environment.

参考译文：电子跃迁最有可能在分子及其环境中原子核位置不变的情况下发生。

➢ At room temperature, most of the molecules are in the lowest vibrational level of the ground state.

参考译文：在室温下，大多数分子处于基态的最低振动能级。

Exercises

1. Discuss the following questions

(1) What is the Franck-Condon principle?

(2) What is the Beer-Lambert law?

(3) Why the total spin quantum number of two electrons with parallel spins is 1?

(4) What could be the cause of bathochromic shift or hypsochromic shift in absorption spectrum?

2. Translate the following into Chinese

(1) The absorption or reflectance in the visible range directly affects the perceived colour of the chemicals involved.

(2) A σ orbital can be formed either from two s atomic orbitals, or from one s and one p atomic orbital, or from two p atomic orbitals having a collinear axis of symmetry.

(3) For example, in ethylene ($CH_2{=}CH_2$), the two carbon atoms are linked by one σ and one π bond. Absorption of a photon of appropriate energy can promote one of the π electrons to an antibonding orbital denoted by π*. The transition is then called π → π*.

(4) The n → π* transition deserves further attention. Upon excitation, an electron is removed from the oxygen atom and goes into the π* orbital localized half on the carbon atom and half on the oxygen atom.

(5) When one of the two electrons of opposite spins (belonging to a molecular orbital of a molecule in the ground state) is promoted to a molecular orbital of higher energy, its spin is in principle unchanged so that the total spin quantum number ($S = \sum S_i = +\frac{1}{2}$ or $-\frac{1}{2}$) remains equal to zero.

(6) The Beer-Lambert law states that the absorbance of a solution is directly proportional to the concentration of the absorbing species in the solution and the path length.

3. Translate the following into English

(1) π轨道是由两个p电子轨道横向重叠形成的。

(2) 分子可以具有如氧或氮等杂原子上的非成键电子。

(3) 根据洪德定则，同一分子的三重态比单线态具有更低的能量。

(4) π 轨道的重叠使电子在整个系统中离域。

(5) 由于环境变化的影响，吸收光谱有可能发生波长红移或蓝移。

(6) 紫外/可见吸收光谱可以用来测定溶液中发色团的浓度。

UNIT 2 Infrared Spectroscopy

1. Discovery of Infrared (IR) Region

The discovery of an invisible component beyond the red end of the solar spectrum [modern meaning—infrared (IR) region in a broad sense] is ascribed to William Herschel, a German-born British astronomer, who is famous for the Herschel telescope. In 1800, he investigated the effect of sunlight divided from violet to red by a prism on temperature increase. He used just sunlight, a prism, and thermometers. Fig.2.8 shows his portrait and the experimental setup he employed. He happened to find that the significant temperature increase occurred even outside of red. He thought there was a different kind of invisible radiation from visible light beyond the red end of sunlight and named this radiation "heat ray". This was really a great discovery in science, but even he could not imagine that this is light. He was 62 years old when he discovered "heat ray". 62 years old in 1800 probably corresponds to today's 80 years old or so. Thus, his discovery demonstrates that even very senior scientist can have intensive serendipity.

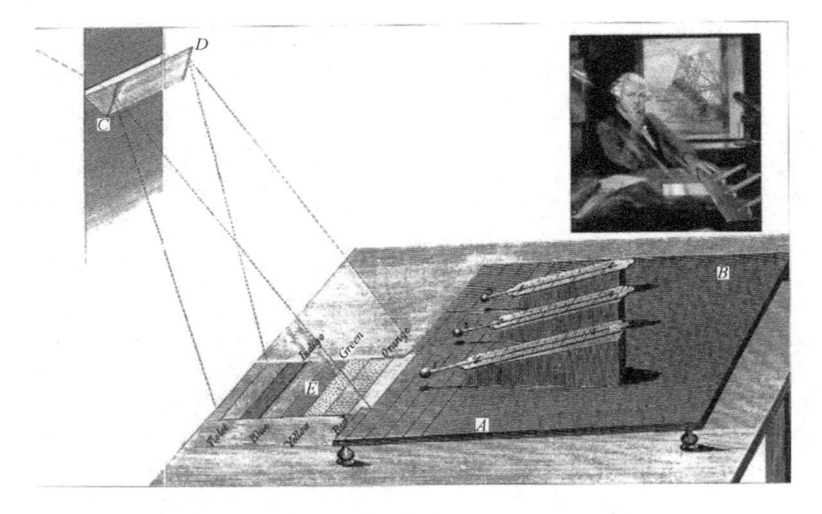

Fig.2.8 The portrait of William Herschel and his experimental setup in 1800

Interestingly enough, just one year after the discovery of "heat ray", Johann Ritter, a German scientist found another invisible component beyond the violet end of the solar spectrum based on an experiment of blackening of silver chloride. In this way, a new era of light was opened at the turning point from the eighteenth century to the nineteenth century. After the discovery of "heat ray", many scientists investigated it. In 1835, it was confirmed that "heat ray" is invisible light which has longer wavelength than visible light. He named this light "infrared (IR)". Maxwell elucidated theoretically in 1864 that ultraviolet, visible, and IR light are all electromagnetic waves. In 1888, he proved it experimentally.

2. Interaction of Light Waves with Molecules: Principles of IR Spectroscopy

When a molecule is irradiated with IR light, it absorbs the light under some conditions. The energy $h\nu$ of the absorbed IR light is equal to an energy difference between a certain energy level of vibration of a molecule (having an energy E_m) and another energy level of vibration of the molecule (having an energy E_n). In the form of an equation

$$h\nu = E_n - E_m \tag{2.2}$$

This equation is known as Bohr frequency condition. In other words, absorption of IR light takes place based on a transition between energy levels of a molecular vibration. Therefore, an IR absorption spectrum is a vibrational spectrum of a molecule.

Note that satisfying Eq.2.2 does not always mean the occurrence of IR absorption. There are transitions which are allowed by a selection rule (i.e., allowed transition) and those which are not allowed by the same rule (i.e., forbidden transition). In general, transitions with a change in the vibrational quantum number by $\pm$ 1 are allowed transitions and other transitions are forbidden transitions under harmonic approximation. This is one of selection rules of IR absorption. Another IR selection rule is defined by the symmetry of a molecule. The latter selection rule is a rule that IR light is absorbed when the electric dipole moment of a molecule varies as a whole in accordance with molecular vibration.

Most molecules are in the ground vibrational state at room temperature, and thus, a transition from the state $\nu'' = 0$ to the state $\nu'' = 1$ (first excited state) is possible. Absorption corresponding to this transition is called the fundamental. Although most bands which are observed in an IR absorption spectrum arise from the fundamental, in some cases, also in the IR spectrum, one can observe bands which correspond to transitions from the state $\nu'' = 0$ to the state $\nu'' = 2, 3\ldots$ They are called first, second, overtones. Bands due to combinations are also observed in the IR spectra. However, since overtones and combinations are forbidden with harmonic oscillator approximation, overtone and combination bands are very weak even in real molecules. Because of anharmonicity, although the intensities are weak, the forbidden bands appear.

3. The Three Infrared Regions of Interest in the Electromagnetic Spectrum

In terms of wavelengths, the three regions in micrometers (μm) are the following: ①NIRS (0.7~2.5μm), ②MIRS (2.5~25μm), ③FIRS (25~300μm). In terms of wavenumbers, the three regions in cm^{-1} are: ①NIRS (14000~4000cm^{-1}), ②MIRS (4000~400cm^{-1}), ③FIRS (400~10cm^{-1}).

The first region (NIRS) allows the study of overtones and harmonic or combination vibrations. The MIRS region is to study the fundamental vibrations and the rotation-vibration structure of small molecules, whereas the FIRS region is for the low heavy atom vibrations (metal-ligand or the lattice vibrations). Infrared (IR) light is electromagnetic (EM) radiation with a wavelength longer than that of visible light: ≤0.7μm. One micrometer (μm) is 10^{-6}m.

Experiments continued with the use of these infrared rays in spectroscopy called infrared spectroscopy, and the first infrared spectrometer was built in 1835. IR spectroscopy expanded rapidly in the study of materials and for the chemical characterization of materials that are in our planet as well as beyond the planets and the stars. The renowned spectroscopists, Hertzberg, Coblenz and Angstrom in the years that followed had advanced the cause of infrared spectroscopy greatly. By 1900, IR spectroscopy became an important tool for identification and characterization of chemical compounds and materials. For example, the carboxylic acids, R—COOH, show two characteristic bands at 1700cm^{-1} and near 3500cm^{-1}, which correspond to the C═O and O—H stretching vibrations of the carboxyl group, —COOH. Ketones, R—CO—R absorb at 1740~1730cm^{-1}. Saturated carboxylic acids absorb at 1710cm^{-1}, whereas saturated/aromatic carboxylic acids absorb at 1690~1680cm^{-1} and carboxylic salts or metal carboxylates absorb at 1610~1550cm^{-1}. By 1950, IR spectroscopy was applied to more complicated molecules such as proteins by Elliot and Ambrose. These studies showed that IR spectroscopy could also be used to study biological molecules, such as proteins, DNA and membranes, in general.

Physicochemical techniques, especially infrared spectroscopic methods, are non-distractive and may be the ones that can extract information concerning molecular structure and characterization of many materials at a variety of levels. Spectroscopic techniques those based upon the interaction of light with matter have for long time been used to study materials both in vivo and in ex vivo or in vitro. Infrared spectroscopy can provide information on isolated materials, biomaterials, such as biopolymers as well as biological materials, connective tissues, single cells and in general biological fluids to give only a few examples. Such varied information may be obtained in a single experiment from very small samples. Clearly then infrared spectroscopy provides information on the energy levels of the molecules in wavenumbers (cm^{-1}) in the region of electromagnetic spectrum by studying the vibrations of the molecules, which are also given in wavelengths (μm).

音频

Infrared spectroscopy is the study of the interaction of matter with light radiation when waves travel through the medium (matter). The waves are electromagnetic in nature and interact with the polarity of the chemical bonds of the molecules. If there is no polarity (dipole moment) in the molecule, the infrared interaction is inactive and the molecule does not produce any IR spectrum.

4. The Techniques of Infrared Spectroscopy

We have two types of IR spectrophotometers: the classical and the Fourier transform spectrophotometers with the interferometer.

The main elements of the standard IR classical instrumentation consist of 4 parts (see Fig.2.9): ①a light source of irradiation, ②a dispersing element, diffraction grating or a prism, ③a detector, ④optical system of mirrors. A schematic diagram of a two-beam absorption spectrometer is shown in Fig.2.9.

The infrared radiation from the source by reflecting to a flat mirror passes through the sample and reference monochromator then through the sample. The beams are reflected on a rotating mirror, which alternates passing the sample and reference beams to the dispersing element and finally to the detector to give the spectrum. As the beams alternate, the mirror rotates slowly and different frequencies of infrared radiation pass to the detector.

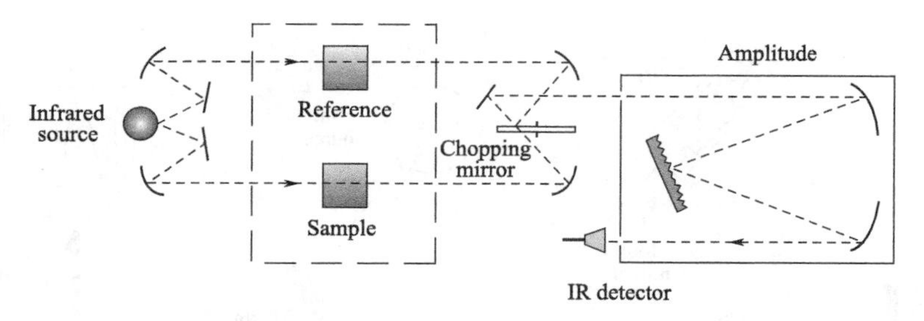

Fig.2.9 A schematic diagram of the classical dispersive IR spectrophotometer

The modern spectrometers came with the development of the high-performance Fourier transform infrared spectroscopy (FT-IR) with the application of a Michelson interferometer. Michelson FT-IR spectrometer has the following main parts: light source, beam splitter (half silvered mirror), translating mirror, detector, optical system (fixed mirror), as shown in Fig.2.10. Both classical and modern IR spectrometers give the same information, the main difference is the use of Michelson interferometer, which allows all the frequencies to reach the detector at once and not one at the time.

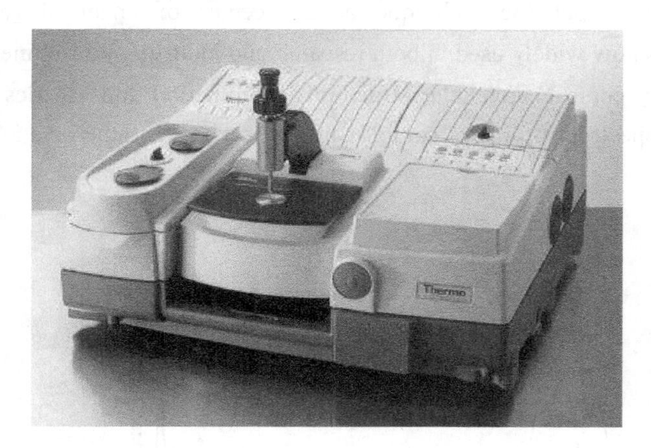

Fig.2.10 Michelson FT-IR spectrometer

In the 1870s, Michelson was measuring light and its speed with great precision and reported the speed of light with the greatest precision to be 299940km·s^{-1}, and he was awarded the Nobel Prize in 1907. However, even though the experiments in interferometry by Michelson and Morley were performed in 1887, the interferograms obtained with this spectrometer were very complex and could not be analyzed at that time because the mathematical formulae of Jean Baptiste Fourier series in 1882 could not be solved. Until the invention of lasers and the high performance of electronic computers, the mathematical formulae of Fourier transform a number of points into waves and finally into the spectra.

The lasers to the Michelson interferometer provided an accurate method (see Fig.2.11) of monitoring displacements of a moving mirror in the interferometer with a high-performance computer, which allowed the complex interferogram analyzed and converted via Fourier transform to give spectra.

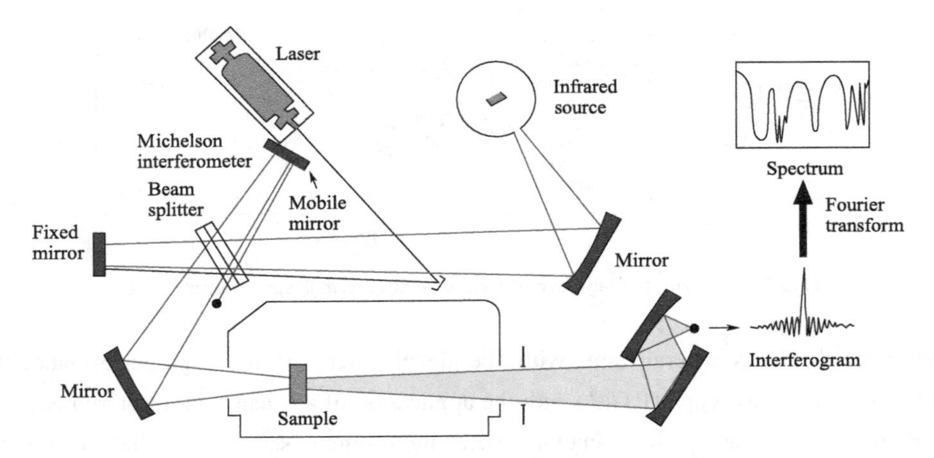

Fig.2.11 Schematic illustration of a modern FT-IR spectrophotometer

Infrared spectroscopy underwent tremendous advances with improvements in instrumentation and electronics, which put the technique at the center of chemical research. The FT-IR spectrophotometry is now widely used in both research and industry as a routine method and reliable technique for quality control, molecular structure determination and kinetics in biosciences (see Fig.2.12). Here the spectrum of a very complex matter, such as an atheromatic plaque is given and interpreted.

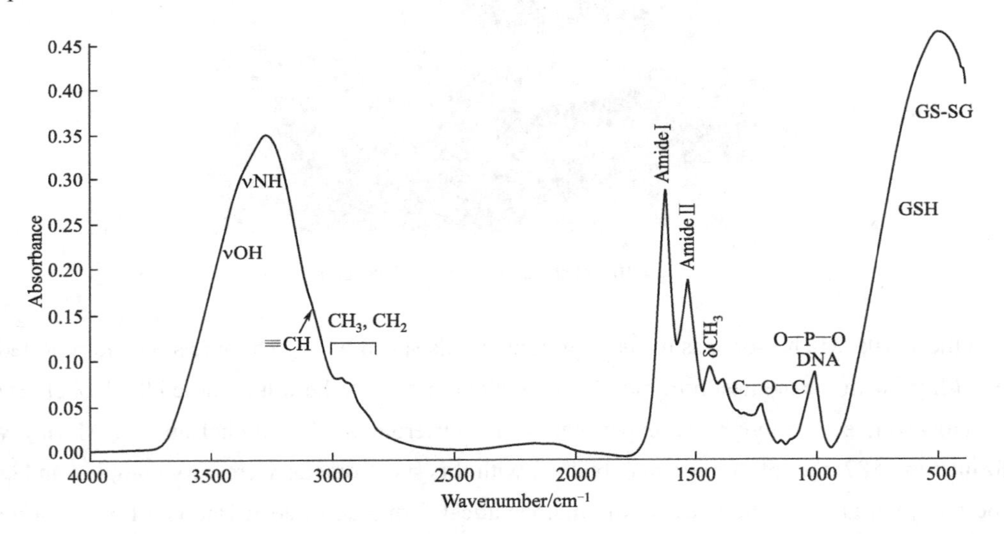

Fig.2.12 FT-IR spectrum of a coronary atheromatic plaque with the characteristic absorption bands of proteins, amide, O—P—O of DNA or phospholipids, disulfide groups, etc.

The addition of a reflecting microscope to the IR spectrometer permits to obtain IR spectra of small molecules, crystals and tissues cells, thus we can apply the IR spectroscopy to biological systems, such as connective tissues, blood samples and bones, in pathology in medicine. In Fig.2.13, it is shown the microscope imaging of cancerous breast tissues and its spectrum.

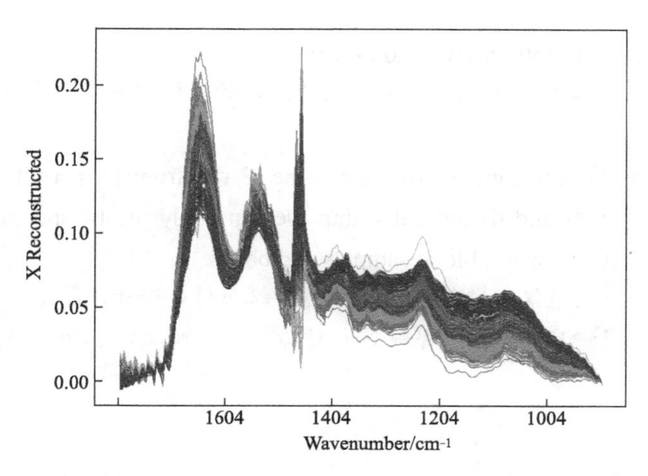

Fig.2.13 Breast tissue: a 3-axis diagram and the mean spectral components

IR spectroscopy is also used in both research and industry for measurement and quality control. The instruments are now small and portable to be transported, even for use in field trials. Samples in solutions can also be measured accurately. The spectra of substances can be compared with a store of thousands of reference spectra.

Words and Expressions

prism *n.* 棱镜
thermometer *n.* 温度计
silver chloride 氯化银
electromagnetic *adj.* 电磁的
irradiate *v.* 照射
vibration *n.* 振动
equation *n.* 方程式
symmetry *n.* 对称
oscillator *n.* 振子
approximation *n.* 近似值
rotation *n.* 旋转
diffraction *n.* 衍射
monochromator *n.* 单色器
kinetics *n.* 动力学
crystal *n.* 结晶
phospholipid *n.* 磷脂
pathology *n.* 病理
atheromatic *adj.* 动脉粥样硬化的

Notes

➢ Infrared spectroscopic methods are non-distractive and may be the ones that can extract information concerning molecular structure and characterization of many materials at a variety of levels.

参考译文：红外光谱技术具有非破坏性，可以在不同层次上提取多种材料的分子结构和表征信息。

➢ Spectroscopic techniques based upon the interaction of light with matter have for long time been used to study materials both in vivo and ex vivo.

参考译文：长期以来，以光与物质相互作用为基础的光谱学技术一直被用于研究生物体内或体外的材料。

➢ The addition of a reflecting microscope to the IR spectrometer permits to obtain IR spectra of small molecules, crystals and tissues cells, thus we can apply the IR spectroscopy to biological systems, such as connective tissues, blood samples and bones.

参考译文：在红外光谱仪上增加反射显微镜可以获得小分子、晶体和组织细胞的红外光谱，因此可以将红外光谱应用于生物系统，如结缔组织、血液样本和骨骼。

Exercises

1. Discuss the following questions

(1) What are the suitable research objects of the NIRS, MIRS, and FIRS region?

(2) Why significant temperature increase occurred even outside of red in William Herschel's experiment?

(3) What is the main difference between the classical and the Fourier transform spectrophotometer?

2. Translate the following into Chinese

(1) He happened to find that the significant temperature increase occurred even outside of red.

(2) In 1835, it was confirmed that "heat ray" is invisible light which has longer wavelength than visible light.

(3) In other words, absorption of IR light takes place based on a transition between energy levels of a molecular vibration.

(4) Another IR selection rule is a selection rule which is defined by the symmetry of a molecule.

(5) Most molecules are in the ground vibrational state at room temperature, and thus, a transition from the state $v'' = 0$ to the state $v'' = 1$ (first excited state) is possible.

(6) Although most bands which are observed in an IR absorption spectrum arise from the fundamental, in some cases, also in the IR spectrum, one can observe bands which correspond to transitions from the state $v'' = 0$ to the state $v'' = 2, 3, \ldots$

3. Translate the following into English

(1) 当一个分子受到红外光照射时，在某些条件下这个分子会对光产生吸收。

(2) 中红外区可以用于研究小分子的振动和键的旋转。

(3) 红外光是一种电磁辐射，其波长大于可见光。

(4) 红外光谱在物质研究和物质化学表征方面发展迅速。

(5) 红外光谱也可以用于研究生物分子，如蛋白质、DNA 和膜。

(6) 红外光谱学研究光波在介质中传播时物质与光辐射的相互作用。

UNIT 3 Nuclear Magnetic Resonance (NMR) Spectroscopy

Nuclear magnetic resonance (NMR) has transformed the research areas of chemistry, biochemistry and medicine, but much of its fundamentals remain obscure for the non-initiated. NMR is a technique based on the absorption of radiofrequency radiation by atomic nuclei in the presence of an external magnetic field. In this unit, we describe the physical basis of phenomena within NMR spectroscopy from both a quantum-theory perspective and a classical view, to provide any prospective user the basic concepts underlying the technique. The spectroscopic notion of energy level population is described, as well as a basic introduction to the theory and mechanisms of spin relaxation. The application of radiofrequency pulses to produce the NMR signal, its conversion to the frequency domain by Fourier transform and the typical instrumental set-up of magnetic resonance are also covered.

NMR is based on the magnetic properties of atomic nuclei, which may be considered to be composed of spinning particles in the simplest model. NMR uses an effect which is well-known in classical physics: When two pendulums are joined by a flexible axle and one of them is forced into oscillation, the other is forced into movement by the common flexible support and the energy will flow between the two. This flow of energy is most efficient when the frequencies of the two movements are identical: the so-called resonance condition. Another example of the resonance condition is found in radio antennas. A radio antenna responds to broadcast radiofrequency signal through the movement of electrons, which shift up and down in the antenna at the same frequency as that of the broadcast signal. In both examples, resonance (the same frequency) is the key, as it is in NMR.

Nuclei are shielded to differ extents by electrons that either reside within orbitals associated directly with a particular atom or within bonding orbitals that are partially de-localized. These shielding effects give rise to the fine structure of an NMR spectrum. It follows that if similar nuclei reside within slightly differing chemical environment, they are shielded by varying amounts, so different magnetic field strengths will be needed to bring these nuclei into resonance. Resonance is achieved by sweeping the intensity of the magnetic field over a very small range to bring each of the nuclei of a given isotopic configuration into resonance.

1. Process of NMR

The process of nuclear magnetic resonance occurs in a spectrometer in the following method: ①Specific magnetic field strengths are generated on the z axis by a powerful magnet; ②A sample is placed in the spectrometer and is bombarded with radiofrequency (RF) at a constant pace along the x axis; ③When the external applied field establishes the correct intensity, resonance occurs as the nuclei of the sample absorb the supplied RF; ④Resonance causes the nuclei to absorb a small current of electricity, which is noted by the receiver coil encompassing the sample; ⑤The spectrometer amplifies the current from the receiver into a display of signals, known as an NMR spectrum. The resulting illustration of an NMR spectrum can be deciphered into the sample's structure (Fig.2.14).

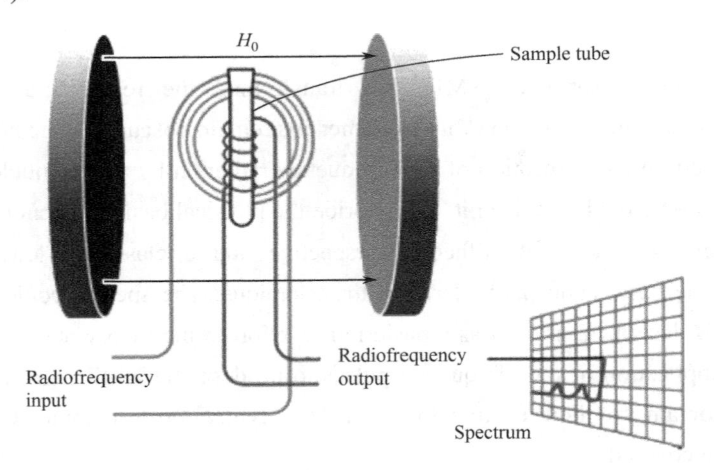

Fig.2.14 Schematic diagram of a nuclear magnetic resonance spectrometer

NMR spectra utilize signals to illustrate different types of H atoms as electronic signals on the recording of spectrum. Each signal contains essential information: the integral, multiplicity, and chemical shift. Multiplicity is a reflection on the number of non-equivalent Hs located on adjacent carbons. Chemical shift provides information about the chemical environment of each H. Integral (integrated area) is the relative number of Hs in the signal. The sum of the integrals can be used to determine the ratios of the types of Hs present in the molecule (Fig.2.15).

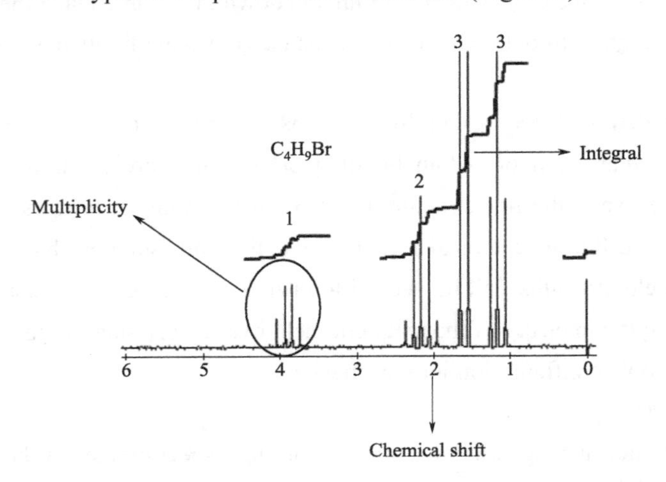

Fig.2.15 ^{1}H-NMR spectrum of C_4H_9Br

2. The Shielding Screening Constant

When any type of nucleus in the sample is subjected to the same external magnetic field, B_0, it will experience the corresponding Larmor frequency, ω. Therefore, the NMR spectrum will be composed of identical resonance lines of the same frequency for a particular type of nucleus. Fortunately for our purposes, this is not the case experimentally and each nucleus resonates with a particular frequency. To show this point, we consider as an example a simple molecule ethanol (CH_3CH_2OH) that would present an "ideal" proton NMR spectrum as shown in Fig.2.16. Three groups of signals can be clearly differentiated, corresponding to the chemical differences of the hydrogen atoms in the molecule. One group of signals corresponds to the methyl group ($—CH_3$), a second one to the methylene protons ($—CH_2—$) and a third for the hydroxyl group ($—OH$). Chemically speaking, it is obvious that each group of hydrogens in the ethanol molecule has different molecular environments, and therefore will experience slightly different local fields. If we compare the frequency for each ^1H nucleus in ethanol with that calculated by γB_0, we could find that each resonance frequency measured experimentally is different from the theoretically predicted one. That is, any of the resonance frequencies observed can be described as

$$\omega = \gamma B_0 - \sigma\gamma B_0 = \gamma B_0(1 - \sigma) \tag{2.3}$$

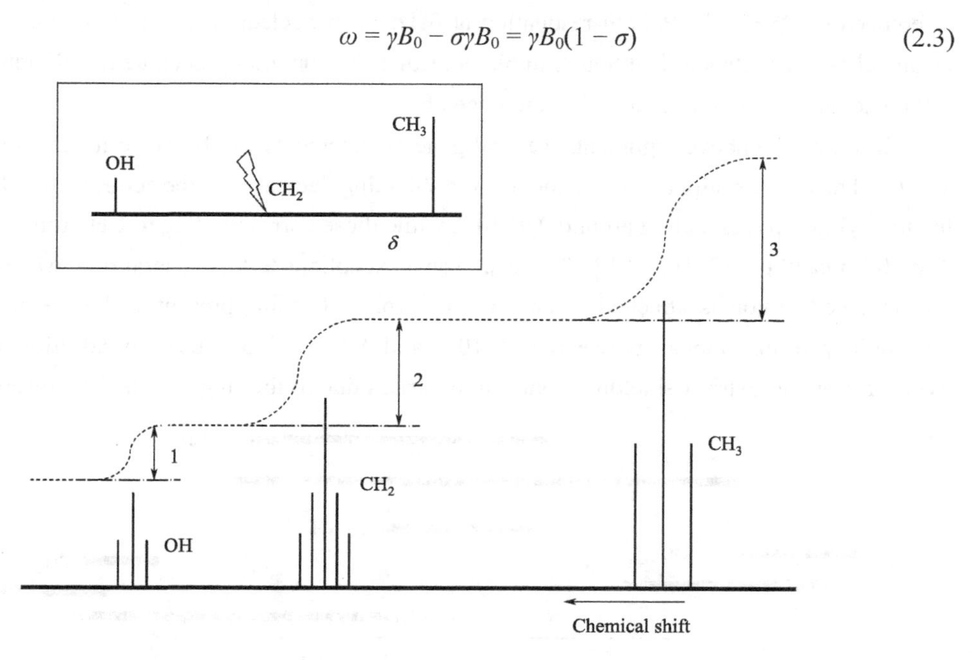

Fig.2.16 NMR spectrum of ethanol

where the factor σ is called the shielding constant. σ can be positive or negative depending on whether the local field adds or subtracts from the applied field. The σ constant varies with B_0 along the axes within the molecule, and can be described mathematically as a tensor. Since σ varies with the chemical environments, identical atoms within a molecule will resonate at different frequencies and therefore render unique NMR signals, giving rise to the richness and power of NMR spectroscopy.

3. The Chemical Shift

It could be seen that different nuclei experience differing shielding effects due to the cloud of electrons that surrounds them. The chemical shift, δ, describes the change in frequency of the magnetic field that must be applied in order to bring specified nuclei into resonance with the applied field. The chemical shift represents the ratio of the change in the magnetic field with respect to the magnetic field required to bring nuclei with no shielding effects into resonance; as such the chemical field is a dimensionless parameter and is normally expressed in terms of 10^{-6} for the required change in the applied field.

To quantify the resonance frequencies, signal reference standards are required. The common standard for protons is the ^1H resonance of $Si(CH_3)_4$, a chemically stable molecule that presents a unique ^1H resonance, which is very intense as it is contributed to by 12 equivalent proton atoms, to which the arbitrary value of 0Hz frequency is assigned. Conveniently, the zero reference frequency for the ^{13}C nucleus is also that of the methyl carbon of TMS. Alternatively, the ^1H signal of the residual non-deuterated solvent in the sample can also be used as a reference, like $CHCl_3$ or DMSO. For the H_2O/D_2O samples typically used in biomolecular NMR, ^1H resonances are normally referenced to DSS or TSP, both resonating at 0Hz. Each nucleus in the periodic table has its own chemical as NMR standard, although in biomolecules the common procedure to calibrate the ^{13}C and ^{15}N nuclei is to take into account the reference of ^1H.

Based on the above arguments, several general statements can be made for interpreting proton spectra. Thus, in the aliphatic C—H bonds, the shielding decreases in the series $CH_3>CH_2>CH$, with the methyl groups resonating around 0.9×10^{-6}, while those corresponding to CH protons lay between 3.5×10^{-6} and 10.0×10^{-6} (Fig.2.17). There are some exceptions to this general rule, which occur when the observed proton is attached to an atom or group of atoms presenting high electronegativity. Aromatic protons appear between 6.0×10^{-6} and 8.0×10^{-6} because, in addition to their sp^2 hybridization, an extra deshielding contribution arises due to the ring-effects. The resonance signals

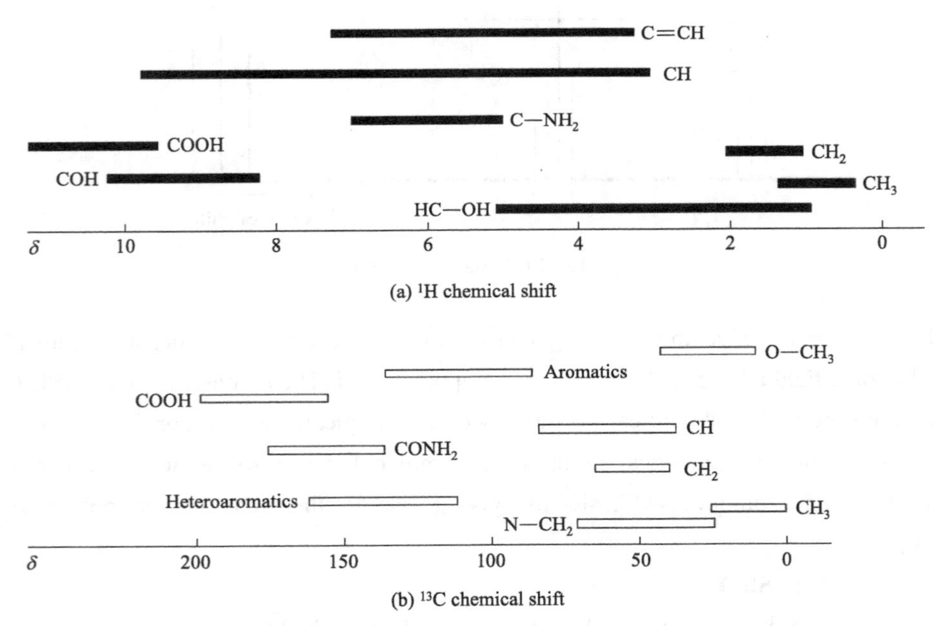

Fig.2.17 Resonance frequency ranges of characteristic chemical functional groups

of protons belonging to aldehydes and carboxylic acids appear at very large δ values ($> 8.5\times10^{-6}$). Carbon atoms bound to nitrogen, oxygen or halogens induce large displacements in the resonance of neighbouring protons leading to deshielded chemical shifts. The NMR signals of labile protons like OH, NH, NH_2 and CO_2H are concentration-, temperature- and solvent-dependent due to possible exchange reactions with the solvent [Fig.2.17(a)].

These chemical shift rules can also be applied to ^{13}C. For instance, the chemical shifts of ^{13}C belonging to aliphatic chains are shielded, and those involved in aldehydes or acids are deshielded. However, since the γ of ^{13}C is different to that of ^{1}H, the range of frequencies/chemical shifts is very different, with ^{13}C δs extending to above 200×10^{-6} [Fig.2.17(b)].

4. Spin-spin Coupling

Spin-spin coupling is a process in which the spinning nuclei of two or more atoms (in different chemical environments) interact to generate a further feature of the fine structure of the NMR spectrum. The effects of spin-spin coupling are seen by the splitting of a peak at a given chemical shift to give a multiplet peak.

The protons in one functional group undergo mutual spin-spin coupling with the protons of an adjacent functional group. There is a simple rule for predicting how spin-spin coupling between adjacent functional groups will split peaks into multiplets. If a functional group, A, with N protons undergoes spin-spin coupling with another (or multiple) proton(s) containing group(s), then the NMR peak due to the protons on neighbouring groups will be split into a multiplet containing N+1 peaks. This rule holds for all functional groups in a molecule, so the effects of all spin-spin interactions by adjacent groups should be considered. If two nuclei reside in equivalent chemical environments, then the peak due to these nuclei will not be split and it follows that spin-spin coupling effects are only seen in NMR spectra if protons residing in differing chemical environments are present.

The N+1 spin-spin coupling rule is a consequence of the number or orientations in which protons of a particular field can align themselves relative to the magnetic field. These effects are, however, evidenced by the splitting of NMR peaks that correspond to the protons of an adjacent proton-containing group. The detailed reasoning of why this occurs is embedded in quantum theory and is beyond the scope of this book. We shall consider how we arrive at N+1 orientations for the alignment of protons with the magnetic field.

As shown in Fig.2.18, the ^{1}H NMR spectrum of CH_3CH_2I, it represents one of the simplest molecules to possess protons in different chemical environments that are capable of undergoing spin-spin coupling with each other. This molecule possesses five protons within its methyl (three) and methylene (two) groups. The ^{1}H NMR spectrum consists of two groups of peaks (the peak at 0 is due to the tetramethylsilane-TMS standard). In this example, the protons of the CH_3 group couple with the protons of the CH_2 group and so the CH_3 peak is split into a triplet since $N+1 = 2+1 = 3$. The CH_2 peak is split into a quartet (i.e., $N+1 = 3+1 = 4$).

Integration curves are also shown in Fig.2.18. The heights of each describe the relative areas under the peaks. We can thus see that the total area for the CH_3 protons is 1.5 times that for CH_2

protons. The areas of these peaks are, therefore, in the ratio 3∶2, reflecting the number of protons in each functional moiety.

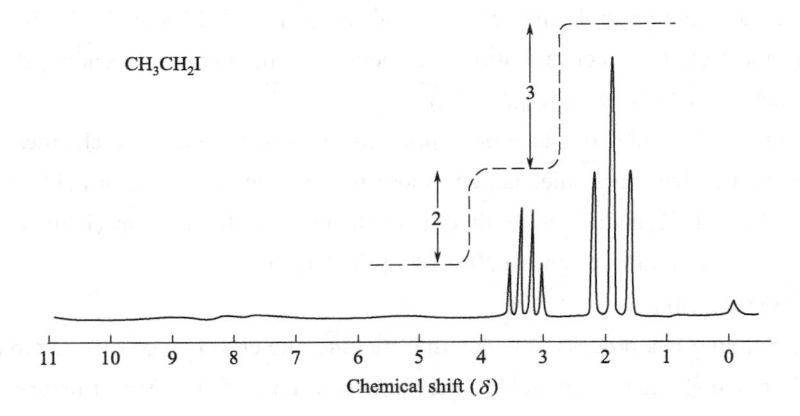

Fig.2.18 NMR spectrum for ethyliodide

Another compound we shall consider is 1, 3-dichloropropane, $ClCH_2CH_2CH_2Cl$, the spectrum of which is shown in Fig.2.19. We can see the protons form two groups, the central methylene groups and the two chloromethyl groups that are identical. The spectrum shows two groups of peaks in the expected 2∶1 ratio. The group centered at 3.9×10^{-6} is due to the chloromethyl groups and is split by the two protons of the methylene group into a triplet. The methylene group of peaks centered at 2.1×10^{-6} is split by four equivalent protons into a quintet.

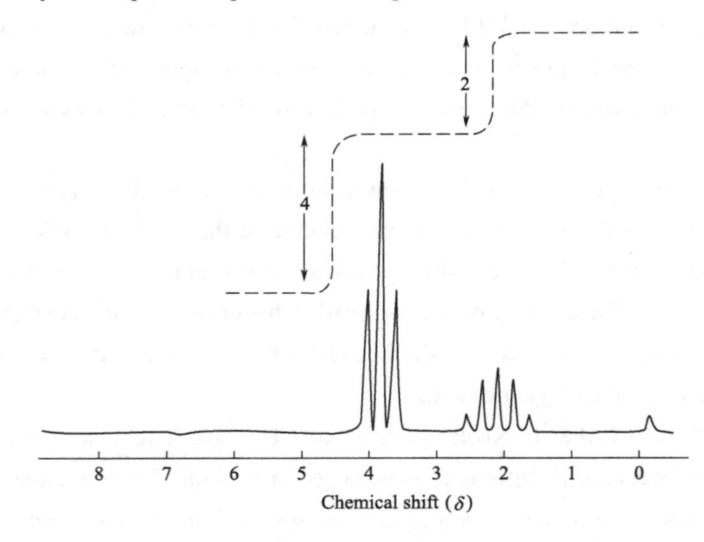

Fig.2.19 NMR spectrum of 1, 3-dichloropropane

Words and Expressions

spectroscopic *adj.* 光谱学的
radiofrequency *n.* 无线电频率
magnetic field 磁场
resonance *n.* 共振
shielding constant 屏蔽常数
chemical environment 化学环境
chemical shift 化学位移
electronegativity *n.* 电负性
carboxylic acid 羧酸
spin-spin coupling 自旋-自旋耦合
orientation *n.* 方向
nuclei *n.* 原子核
tetramethylsilane *n.* 四甲基硅烷
1, 3-dichloropropane *n.* 二氯丙烷
establish *v.* 建立
multiplet *n.* 多重态
proton *n.* 质子
aliphatic chain 脂肪链
chloromethyl *n.* 氯甲基

Notes

➢ The chemical shift, δ, describes the change in frequency of the magnetic field that must be applied in order to bring specified nuclei into resonance with the applied field.

参考译文：化学位移δ描述了为了使特定原子核与所施加的电场共振而必须施加的磁场频率的变化。

➢ Spin-spin coupling is a process in which the spinning nuclei of two or more atoms (in different chemical environments) interact to generate a further feature of the fine structure of the NMR spectrum.

参考译文：自旋-自旋耦合是以两个或多个原子的自旋核（在不同的化学环境中）相互作用以产生 NMR 光谱精细结构的进一步特征的过程。

➢ The protons in one functional group undergo mutual spin-spin coupling with the protons of an adjacent functional group.

参考译文：一个官能团中的质子与相邻官能团的质子发生自旋-自旋耦合。

Exercises

1. Discuss the following questions

(1) What is shielding screening constant?

(2) What does the chemical shift describe?

(3) What is *N*+1 spin-spin coupling rule?

(4) What is spin-spin coupling?

2. Translate the following into Chinese

(1) Nuclear magnetic resonance (NMR) has transformed the research areas of chemistry, biochemistry and medicine, but much of its fundamentals remain obscure for the non-initiated.

(2) If two nuclei reside in equivalent chemical environments, then the peak due to these nuclei will not be split and it follows that spin-spin coupling effects are only seen in NMR spectra if protons residing in differing chemical environments are present.

(3) The protons in one functional group undergo mutual spin-spin coupling with the protons of an adjacent functional group.

(4) Aromatic protons appear between 6×10^{-6} and 8×10^{-6} because, in addition to their sp^2 hybridization, an extra deshielding contribution arises due to the ring-effects.

(5) Spin-spin coupling is a process in which the spinning nuclei of two or more atoms (in different chemical environments) interact to generate a further feature of the fine structure of the NMR spectrum.

3. Translate the following into English

(1) NMR 基于原子核的磁性，在最简单的模型中可以认为它是由旋转粒子组成的。

(2) 屏蔽作用产生了 NMR 光谱的精细结构。

(3) 自旋-自旋耦合的效果可通过将给定化学位移处的峰拆分以得到多重峰来评价。

(4) 由于 σ 随化学环境而变化，分子内相同的原子会在不同的频率共振，从而呈现独特的 NMR 信号，提高 NMR 光谱的多样性和功能。

(5) 当样品中任何类型的原子核受到相同的外部磁场 B_0 作用时，都会经历相应的拉莫尔频率 ω。

UNIT 4 Mass Spectrometry

A mass spectrometer is a device for producing and weighing ions from a compound for which we wish to obtain molecular weight and structural information (its schematic diagram is shown in Fig.2.20). All mass spectrometers use three basic steps: molecules M are taken into the gas phase; ions, such as the cations $M^{\bullet+}$, MH^+ or $M+Na^+$, are produced from them; and the ions are separated according to their mass-to-charge ratios (m/z). The value of z is normally one, and since e is a constant (the charge on one electron), m/z gives the mass of the ion. Some of the devices that are used to produce gas-phase ions put enough vibrational energy into the ions to cause them to fragment in various ways to produce new ions with smaller m/z ratios. Through this fragmentation, structural information can be obtained.

A mass spectrometer detects only the charged components (e.g., MH^+, and its associated fragments A^+, B^+, C^+, etc.), because only they are contained, accelerated or deflected by the electromagnetic or electrostatic fields used in the various analytical systems, and only they give an electric signal when they hit the collector plate. When the array of ions has been separated and recorded, the output is known as a mass spectrum. It is a record of the abundance of each ion reaching the detector (plotted vertically) against its m/z value (plotted horizontally). The mass spectrum results from a series of competing and consecutive unimolecular reactions, and what it looks like is determined by the chemical structure and reactivity of the sample molecules.

Mass spectrometry is the most sensitive among all these methods. It can be carried out routinely with a few micrograms of sample, and in favorable cases even with picograms (10^{-12}g), making it especially important in solving problems where only a very small sample is available, as in the detection and analysis of such trace materials as pheromones, atmospheric pollutants, pesticide residues, drugs and their metabolites, especially in forensic science and in medical research. It is even possible to use mass spectrometry to analyze individual slices of tissue to detect the change from cancerous to non-cancerous as an operation is in progress.

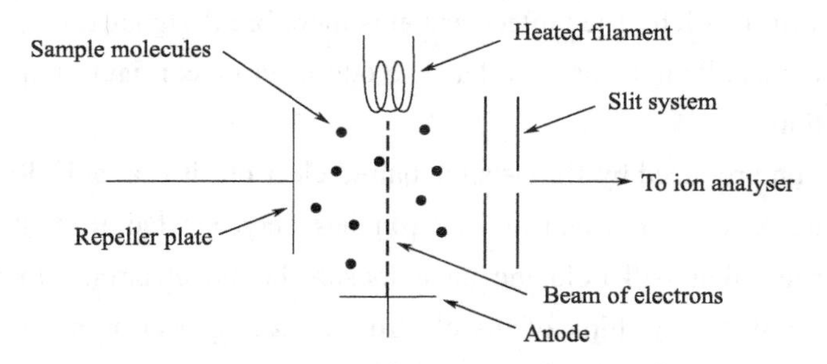

Fig.2.20 Schematic diagram of mass spectrometer

1. Ionization

In a mass spectrometer used for organic analysis, we are usually dealing with the analysis of positive ions derived from molecules rather than atoms. This is carried out by using a variety of methods. Simple ionization of a molecule is shown below in the following equation

$$M: \longrightarrow M^{\bullet+} + e$$

The ion $M^{\bullet+}$ is known as the molecular ion. As mentioned previously, the ion's mass (*m*) and charge (*z*) are measured by the mass spectrometer and it is this ratio that is important. Often ions with only a single positive charge are formed, and thus the mass to charge ratio is "+". This is therefore equal to the mass of the ion itself. The mass of the ion is, of course, related to the relative molecular mass of the molecule.

音频

Thus, all atoms are assigned masses relative to ^{12}C. Fortunately, many elements have average relative atomic masses (sometimes called average atomic weight) which are very close to integral numbers (in amu.). This is a result of the abundances of the isotopes of these elements. This means that when we are roughly calculating molecular masses, we can assign integral values to most of the common elements present in organic molecules.

The relative molecular mass is thus the sum of the average relative atomic masses of the constituent atoms in a compound. As an example, the average relative atomic mass of oxygen is 15.994 (very close to 16), as a result of the abundance of the ^{16}O isotope (99.756%). Similarly, carbon has an average atomic weight of 12.011, as a result of the abundance of the ^{12}C isotope (98.90%). This abundance has valuable ramifications for mass spectrum identification.

Since an electron has a very small mass indeed, i.e., 5.49×10^{-4} of the mass of a proton, the mass loss in the atom is negligibly small upon ionization, when compared to the (mass) resolution of most mass spectrometers. Therefore, we can assume for the purpose of identification that the mass of a molecular ion is equivalent to the mass of the parent molecule from which it is derived.

You may notice in the equation that there are other labels used with molecule and molecular ion, namely the symbols ":" and "$^{\bullet+}$". They are used to help us remember the electron configuration of the molecule. Virtually all organic molecules have an even number of electrons, so the designation in this case is M:. These electrons occupy bonding or non-bonding orbitals. Thus, if upon ionization one electron is removed, then the resultant molecular ion will have an odd electron configuration, i.e., one unpaired electron is left. The molecular ion is therefore designated $M^{\bullet+}$. In other words, it is a radical cation. Note that all molecular ions have an odd-electron configuration.

2. Fragmentation

The molecular ion produced by EI has an unpaired electron (it is a radical-cation, $M^{\bullet+}$). We can think of its structure as one in which an electron has been ejected from the highest occupied molecular orbital, since that will hold the most loosely bound electrons. In a hydrocarbon like *n*-nonane, we can simplistically think of the electron as having been removed from one of the σ bonds. Thus, given sufficient vibrational energy, a bond like the one illustrated in the representation **1** can break with the single electron remaining in the bond moving to the right to give an ion **2** and a radical **3**. Equally plausible would be for it to move in the other direction to give the radical **4** and the ion **5** (Fig.2.21).

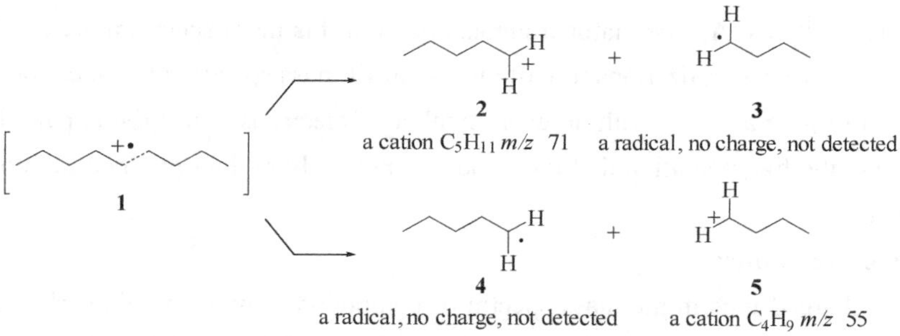

2
a cation C_5H_{11} m/z 71

3
a radical, no charge, not detected

1

4
a radical, no charge, not detected

5
a cation C_4H_9 m/z 55

Fig.2.21 Proposed fragmentation of mass spectrum

The ion **2** with a molecular formula of $C_5H_{11}^+$ gives rise to the peak at m/z 71 and the ion **5** with a molecular formula $C_4H_9^+$ gives rise to the peak at m/z 55. The radicals **3** and **4** with molecular formulae of C_4H_9• and C_5H_{11}• have no charge, they are not detected, and they do not appear in the spectrum. In this way, several C_nH_{2n+1} cations can be generated, giving the ion series m/z 99, 85, 71, 57 and 43. The lower mass ions of this series may be formed, not only directly, but also by the loss of ethylene **7** from one of the ions with higher mass, as in the fragmentation of the pentyl cation **2** to give the propyl cation **6** in a retro-Friedel-Crafts reaction. Thus the ions with the lower m/z values are more abundant than the ions with higher m/z values (Fig.2.22).

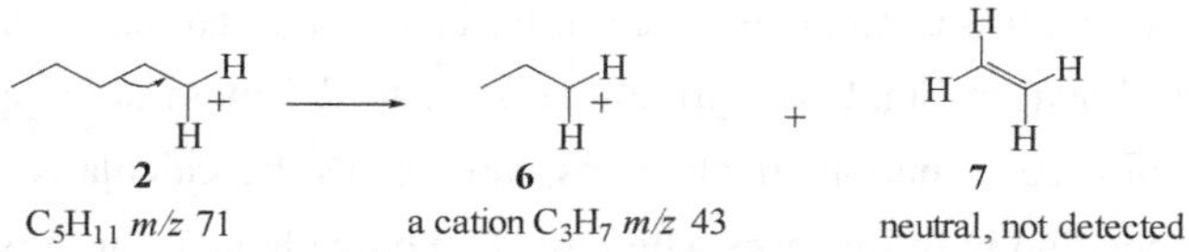

2
C_5H_{11} m/z 71

6
a cation C_3H_7 m/z 43

7
neutral, not detected

Fig.2.22 Proposed fragmentation of C_5H_{11}

A useful way of conveying the structural features responsible for all the major ions is shown in the drawing **8**, where the wavy lines identify the bonds broken and the numbers (and the side on which they are placed) identify the m/z values of the cations produced (Fig.2.23). Note that the fragmentation of any of the C—H bonds is unfavourable, because hydrogen atoms are very high in energy, and the ion at m/z 113 is weak, because the methyl radical is not well stabilized.

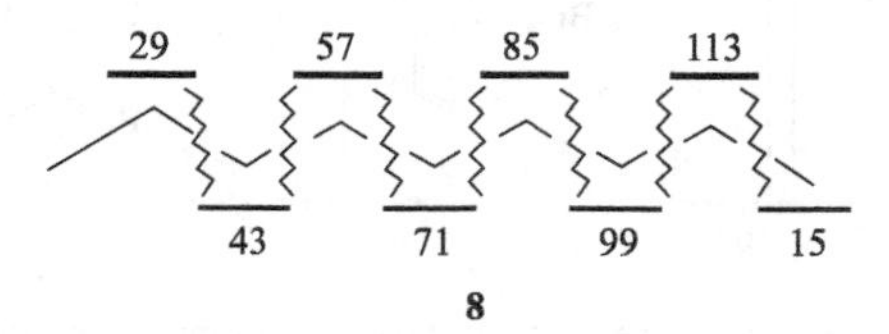

Fig.2.23 Proposed fragmentation

Ions of the general formula $C_nH_{2n-1}^+$ form a minor series of fragment ions at m/z 27, 41 and 55, two mass units below the more prominent ions for the ethyl, propyl and butyl cations at 29, 43 and 57. Their formation occurs by loss of a saturated hydrocarbon molecule or, less commonly, H_2 from ions of the $C_nH_{2n-1}^+$ series. All the major fragmentations in this mass spectrum are nicely explained. They illustrate the two generalizations true of almost all EI mass spectra: radical cations fragment by loss of a radical to give a cation with an even number of electrons. Once the unpaired electron has left the molecule, the fragmentation that takes place is normally by loss of a neutral molecule, rather than of a radical.

3. Isotopic Abundances

All singly charged ions in the mass spectra of compounds which contain carbon give rise to a peak at one mass unit higher, because the natural abundance of ^{13}C is 1.1%. For an ion containing *n* carbon atoms, the abundance of the isotope peak is $n \times 1.1\%$ of the ^{12}C-containing peak. Thus, the molecular ion for nonane ($C_9H_{20}^{\cdot+}$) gives an isotope peak at 129, one mass unit higher than the molecular ion, with an approximate abundance of 10% ($9 \times 1.1\%$) of the abundance at *m*/*z* 128. Obviously, larger molecules with a lot of carbon atoms give much more prominent ($M^{\cdot+}$+1) ions. In the case of small molecules, the probability of finding two ^{13}C atoms in an ion is low, and ($M^{\cdot+}$+2) peaks are accordingly of insignificant abundance. Conversely, $M^{\cdot+}$+2, and even $M^{\cdot+}$+3, $M^{\cdot+}$+4, etc. peaks do become important in very large molecules (see later for details). The ratio of the two peaks ($M^{\cdot+}$+1 : $M^{\cdot+}$) gives a rough measure of the number of carbon atoms in the molecule.

While iodine and fluorine are monoisotopic, chlorine consists of ^{35}Cl and ^{37}Cl in a ratio of approximately 3 : 1, and bromine of ^{79}Br and ^{81}Br in a ratio of approximately 1 : 1. Molecular ions (or fragment ions) containing various numbers of chlorine or bromine atoms therefore give rise to the characteristic patterns shown in Fig.2.24, with all peaks spaced 2 mass units apart. Isotope patterns to be expected from any combination of elements can readily be calculated, and they provide a useful test of ion composition in those cases where polyisotopic elements are involved.

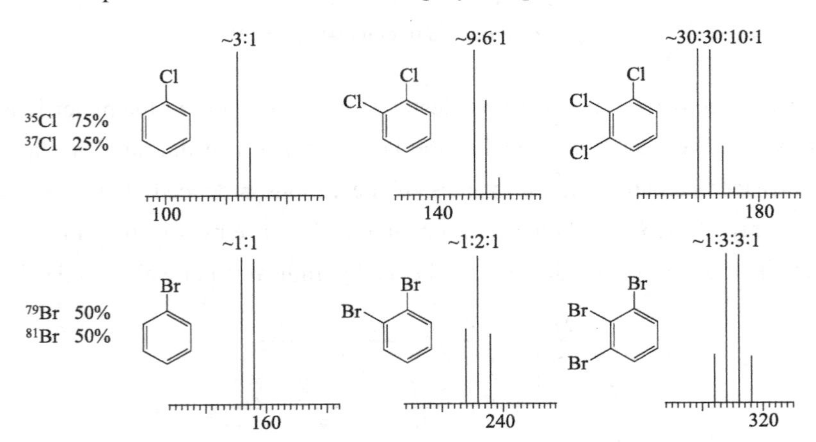

Fig.2.24 Molecular ions in the mass spectra of halogen-containing compounds

4. The Information of an EI Spectrum

Fig.2.25 shows a representative mass spectrum illustrating the main features to be found in every mass spectrum. It is the mass spectrum of *n*-nonane, which proves to have a largely explicable fragmentation pattern.

The most intense peak, at *m/z* 43, is called the base peak. The peak at *m/z* 128 is called the molecular ion $M^{\bullet+}$, with an intensity of 8% of the intensity of the base peak. There are several fragment ions, the intensities of which are similarly recorded as percentages of the height of the base peak. Fragmentation gives rise to a pattern of fragment ions like a fingerprint. A compound may be quickly identified by this pattern if it has already had its mass spectrum recorded. The mass spectrum of nonane would be recorded for publication thus, picking out only the more conspicuous peaks.

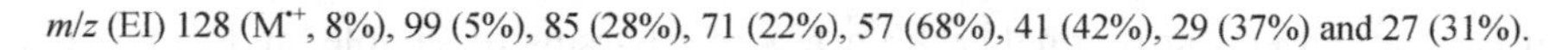
m/z (EI) 128 ($M^{\bullet+}$, 8%), 99 (5%), 85 (28%), 71 (22%), 57 (68%), 41 (42%), 29 (37%) and 27 (31%).

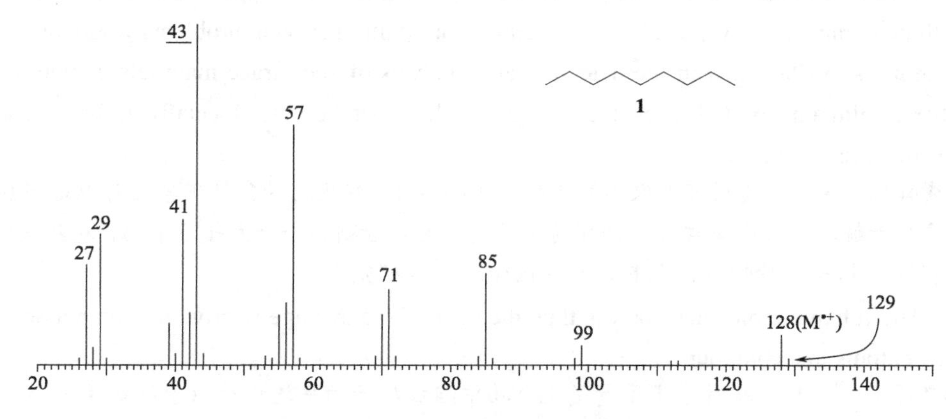

Fig.2.25 EI mass spectrum of *n*-nonane

Words and Expressions

fragment *n.* 碎片
device *n.* 设备; 装置
investigation *n.* 调查
abundance *n.* 丰度
microgram *n.* 微克
atmospheric pollutant 大气污染物
mass spectrometer 质谱仪
isotope *n.* 同位素
ramification *n.* 分支
unpaired electron 未成对电子
ionization *n.* 电离作用
radical cation 自由基阳离子
negligibly *adv.* 可忽视地
electron configuration 电子构型
available *adj.* 可获得的
fragmentation *n.* 分裂
vibrational *adj.* 振动的
fluorine *n.* 氟
bromine *n.* 溴
hydrogen atom 氢原子
mass to charge ratio 质荷比

Notes

➢ The mass spectrum results from a series of competing and consecutive unimolecular reactions, and what it looks like is determined by the chemical structure and reactivity of the sample molecules.

参考译文：质谱是一系列竞争和连续单分子反应的结果，由样品分子的化学结构和反应性质决定。

➢ While iodine and fluorine are monoisotopic, chlorine consists of ^{35}Cl and ^{37}Cl in a ratio of approximately 3∶1, and bromine of ^{79}Br and ^{81}Br in a ratio of approximately 1∶1.

参考译文：碘和氟都只有一种同位素，氯由 ^{35}Cl 和 ^{37}Cl 以约 3∶1 的比例组成，而 ^{79}Br 和 ^{81}Br 的比例约为 1∶1。

➢ It can be carried out routinely with a few micrograms of sample, and in favourable cases even with picograms (10^{-12}g), making it especially important in solving problems where only a very small sample is available, as in the detection and analysis of such trace materials as pheromones, atmospheric pollutants, pesticide residues, drugs and their metabolites, especially in forensic science and in medical research.

参考译文：常规下质谱可以仅用微克数量级的样品，在优化的条件下甚至可以进行皮克级测试，这对于解决仅有少量样品的问题特别重要，例如检测和分析大气污染物、农药残留、药物及其代谢产物等微量物质，尤其是在法医学和医学研究中。

➢ The relative molecular mass is thus the sum of the average relative atomic masses of the constituent atoms in a compound.

参考译文：因此，相对分子质量是化合物中组成原子的平均相对原子质量的总和。

➢ The molecular ion produced by EI has an unpaired electron (it is a radical-cation, $M^{\bullet+}$). We can think of its structure as one in which an electron has been ejected from the highest occupied molecular orbital, since that will hold the most loosely bound electrons.

参考译文：EI 模式产生的离子峰带有一个未成对的电子（它是一个自由基阳离子，$M^{\bullet+}$）。我们可以把它看作从最高占有分子轨道发生一个电子跃迁后的结构，因为该结构能保持分子对电子的最松散束缚。

Exercises

1. Discuss the following question

What is mass spectrometer?

2. Translate the following into Chinese

(1) Because the molecular weight and the molecular formula of an unknown are usually the first pieces of information to be sought in the investigation of a chemical structure, mass spectrometry is often the first of the spectroscopic techniques to be called upon.

(2) The mass spectrum results from a series of competing and consecutive unimolecular reactions, and what it looks like is determined by the chemical structure and reactivity of the sample molecules.

(3) Radical cations fragment by loss of a radical to give a cation with an even number of electrons.

(4) Once the unpaired electron has left the molecule, the fragmentation that takes place is normally by loss of a neutral molecule, rather than of a radical.

(5) All singly charged ions in the mass spectra of compounds which contain carbon also give rise to a peak at one mass unit higher, because the natural abundance of ^{13}C is 1.1%.

(6) While iodine and fluorine are monoisotopic, chlorine consists of ^{35}Cl and ^{37}Cl in a ratio of approximately 3 : 1, and bromine of ^{79}Br and ^{81}Br in a ratio of approximately 1 : 1.

3. Translate the following into English

(1) 质谱仪是一种用于从化合物中产生离子并对其进行称重的设备，我们希望获得该化合物的分子量和结构信息。

(2) 使用质谱分析组织的各个切片，以检测正在进行的手术中患者组织从癌变到非癌的变化界限甚至是有可能的。

(3) 可以容易地计算出元素的任何组合所预期的同位素模式，在涉及多同位素元素的情况下，它们为离子组成提供了有用的测试。

(4) 该化合物在几种条件下的 EI-MS 中表现出几乎相同的行为。

PART 3 Organic Synthetic Processes

UNIT 1 General Introduction of Organic Synthesis

It is important to give some thought to the work up of the reaction before you attempt it. Several aspects which need to be considered are dealt with here. First of all, do make sure that the reaction has indeed finished (by careful analysis using your chosen monitoring system). When using TLC analysis, it is sometimes difficult to judge by spotting the reaction mixture directly on to the TLC plate. In these cases, it is often possible to get a more accurate assessment by withdrawing a small aliquot of reaction mixture by syringe and adding it to a small vial containing a few drops each of ether and aqueous ammonium chloride. Agitation followed by TLC of the organic phase will often give clean, reliable TLC information. This technique can also be used to screen alternative work up conditions, for example adding to water or aqueous base rather than ammonium chloride solution, or using other organic solvents in place of ether.

Having satisfied yourself that the reaction has run to completion, or that it is time to end the experiment, the appropriate "quench" is added to the reaction mixture. Choice of this reagent can be very important in determining the yield of desired product, and it is obviously vital to use a reagent or procedure which is safe. Given that the product is expected to be reasonably stable, which usually is the case, then the choice of procedure for quenching the reaction is determined by the reagent(s) used in the reaction.

Preparative organic reactions rarely give a complete conversion of the starting material into the required product. In a good "synthetic" reaction, a yield of product of 70%~80% or more would be expected, but in addition there may be small amounts of other organic materials formed as by-products. In some cases, reactions also produce polymeric "tarry" material, which may be brown or yellow. Many reactions also produce equimolar amounts of inorganic products, e.g., the metal halides produced in a typical ether synthesis.

The mixture obtained after a reaction has been carried out can be quite complex even for a good synthetic reaction. It will consist of the reaction solvent, the major product, by-product, unreacted starting material, possibly polymeric material and possibly an inorganic product. Obviously, the objective is to separate out the major product in as high a yield as possible and in as pure a state as possible.

There are some reactions (unfortunately very few) in which the major product simply crystallizes out when you cool the reaction mixture down to room temperature. However, the procedure required is usually more complex and involves the successive application of several different techniques. This is generally called "working up" the reaction mixture. The particular methods required will depend on the nature of the mixture and on the chemical and physical properties of the compounds concerned. In straightforward reactions with one major product, the sequence of operations is usually designed to remove the reaction solvent and the inorganic products first, and then to separate the major organic product from the other organic materials present by recrystallization (for solids) or distillation (for liquids). In more complex reactions there may be several "major" products and it may be necessary to separate them by some form of chromatography.

In most teaching experiments and books on synthetic chemistry, the work-up procedure is usually specified in some detail, and even in the descriptions of new experimental work in chemical journals, it is usual to find a description of how the reaction was worked up and the products were isolated. Therefore, it is rare for those under training to have to decide how to work up a reaction, but it is important at all stages to make sure you understand why a particular operation is being carried out and what it is meant to achieve. This will prepare you for the time when you may be doing project work where the chemistry will be new, and for each reaction you will have to work out the best way to get the products separated and purified. You need to call on the wide range of practical skills covered in the following sections, and to use your experience to decide which ones are likely to be most effective for the reaction mixture in question.

Words and Expressions

thin-layer chromatography (TLC)　薄层色谱
withdraw　*v.* 提取
a small aliquot of　一小部分××试样
syringe　*n.* 注射器
vial　*n.* 小瓶
ammonium chloride　氯化铵
quench　*v.* 淬灭
reagent　*n.* 试剂
equimolar　*adj.* 等摩尔的
by-product　*n.* 副产物
polymeric　*adj.* 聚合的
tarry　*adj.* 油状的
recrystallization　*n.* 重结晶
distillation　*n.* 蒸馏
chromatography　*n.* 色谱法

Notes

➢ In these cases, it is often possible to get a more accurate assessment by withdrawing a small aliquot of reaction mixture by syringe and adding it to a small vial containing a few drops each of ether and aqueous ammonium chloride.

参考译文：在这些情况下，通过注射器从反应混合物中取少量试样，加到含有几滴乙醚和氯化铵水溶液的小瓶中，往往会得到更准确的结果。

➢ Choice of this reagent (quench) can be very important in determining the yield of desired product, and it is obviously vital to use a reagent or procedure which is safe.

参考译文：（淬灭）试剂的选择对于判断目的产物的产率是非常重要的，并且使用安全的试剂或实验程序是至关重要的。

➢ In straightforward reactions with one major product, the sequence of operations is usually designed to remove the reaction solvent and the inorganic products first, and then to separate the major organic product from the other organic materials present by recrystallization (for solids) or distillation (for liquids).

参考译文：对于只有一种主产物的一步反应来说，操作顺序通常是这样设计的，首先是去除溶剂和无机产物，然后是通过重结晶（对固体来说）或者蒸馏（对液体来说）从有机产物中分离主要有机产物。

Exercises

1. Discuss the following questions

(1) How to decide whether the reaction has finished?

(2) Why working up is necessary to obtain the desired product?

2. Translate the following into Chinese

(1) Given that the product is expected to be reasonably stable, which usually is the case, then the choice of procedure for quenching the reaction is determined by the reagent(s) used in the reaction.

(2) It will consist of the reaction solvent, the major product, by-product, unreacted starting material, possibly polymeric material and possibly an inorganic product.

(3) Therefore, it is rare for those under training to have to decide how to work up a reaction, but it is important at all stages to make sure you understand why a particular operation is being carried out and what it is meant to achieve.

3. Put the following into English

注射器　　淬灭　　重结晶　　蒸馏　　色谱法

4. Translate the following into English

(1) 在某些情况下，反应还会产生聚合的“焦油状”物质，这些物质可能呈棕色或黄色。

(2) 具体方法的选择取决于混合物的性质以及相关化合物的化学和物理特性。

(3) 在更复杂的反应中，会有多种主产物存在，可能有必要用色谱法分离它们。

UNIT 2 Recrystallization

Purification of a solid by recrystallization from a solvent depends upon the fact that different substances are soluble to different extents in various solvents. In the simplest case, all the unwanted materials are much more soluble than the desired compound. In this case, the sample is dissolved in just enough of the hot solvent to form a saturated solution, the solution is cooled, and the crystals, which will be separated upon cooling, are collected by suction filtration. The soluble impurities remain in solution after cooling and pass through the filter paper with the solvent upon suction filtration.

1. Select a Suitable Solvent

Find a suitable solvent by carrying out small scale tests. Remember that "like dissolves like". The most commonly used solvents in order of increasing polarity are petroleum ether, toluene, chloroform, acetone, ethyl acetate, ethanol, and water. Chloroform and dichloromethane are rarely useful on their own because they are good solvents for the great majority of organic compounds. It is preferable to use a solvent with a boiling point (b.p.) in excess of 60℃, but the b.p. should be at least 10℃ lower than the melting point (m.p.) of the compound to be crystallized, in order to prevent the solute from "oiling out" of solution. In many cases a mixed solvent must be used, and combinations of toluene, chloroform, or ethyl acetate, with the petroleum ether fraction of similar boiling point are particularly useful.

2. Dissolve the Compound in the Minimum Volume of Hot Solvent

Place the crude compound (always keep a few "seed" crystals) in a round-bottom flask fitted with a reflux condenser, add boiling chips and a small portion of solvent, and heat in a water bath (Fig.3.1). Continue to add portions of solvent at intervals until all of the crude has dissolved in the hot refluxing solvent. If you are using a mixed solvent, dissolve the crude in a small volume of the good solvent, heat to reflux, add the poor solvent in portions until the compound just begins to precipitate (cloudiness), add a few drops of the good solvent to redissolve the compound, and allow to cool. When adding the solvent, it is very easy to be misled into adding far too much if the crude is contaminated with an insoluble material, such as silica or magnesium sulphate.

3. Filter the Hot Solution to Remove Insoluble Impurities

This step is often problematic and should not be carried out unless an unacceptable (use your judgement) amount of insoluble material is suspended in the solution. The difficulty here is that the compound tends to crystallize during the filtration so an excess of solvent (ca. 5%) should be added, and the apparatus used for the filtration should be preheated to about the boiling point of the solvent. Use a clean sintered funnel of porosity, or a Hirsch or Buchner funnel, and use the minimum suction needed to draw the solution rapidly through the funnel.

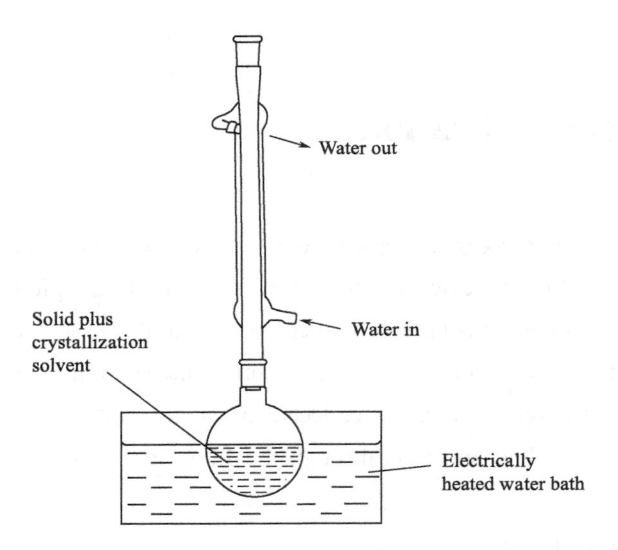

Fig.3.1 Apparatus for recrystallization

If the solution is very dark and/or contains small amounts of tarry impurities, allow it to cool for a few moments, add ca. 2% by weight of decolorizing charcoal, reflux for a few minutes, and filter off the charcoal. Charcoal is very finely divided so it is essential to put a 1cm layer of a filter aid such as celite on the funnel before filtering the suspension. Observe the usual precautions for preventing crystallization in the funnel. Very dark or tarry products should be chromatographed through a short (2~3cm) plug of silica before attempted recrystallization.

4. Allow the Solution to Cool and Form the Crystals

This is usually straightforward except when the material is very impure or as a low m.p. (< 40℃) in which case it sometimes precipitates as an oil. If oil forms, it is best to reheat the solution and then to allow it to cool slowly. Try scratching the flask with a glass rod or adding a few "seed" crystals to induce crystallization, and if this fails try adding some more solvent so that precipitation occurs at a lower temperature. If nothing at all precipitates from the solution, try scratching with a glass rod, seeding, or cooling the solution in ice-water. If all these fails, stop the flask and set it aside for a few days, patience is sometimes the best policy.

5. Filter and Dry the Crystals

When crystallization appears to be complete, filter off the crystals using an appropriately sized sintered glass funnel, it is very important to wash the crystals carefully. As soon as all of the mother liquor has drained through the funnel, remove the suction and pour some cold solvent over the crystals, stir them if necessary, in order to ensure that they are thoroughly washed. Drain off the washing under suction and repeat once or twice more.

After careful washing, allow the crystals to dry briefly in the air and then remove the last traces of solvent under vacuum, in a vacuum oven, in a drying pistol, or on a vacuum line. Take care to protect your crystals against accidental spillage or contamination. If they are placed in a dish, a beaker, or a sample vial, cover with aluminium foil, secure with wire or an elastic band, and punch a few small holes in the foil. If the crystals are in a flask connected directly to a vacuum line, use a tubing adapter with a tap and put a plug of glass wool in the upper neck of the adapter so that the crystals are not blown about or contaminated with rubbish from the tubing, when the air/inert gas is allowed in.

If a relatively high boiling solvent such as toluene is used for the recrystallization, it is essential to heat the sample under vacuum for several hours to ensure that all of the solvent is removed.

6. Recrystallization of 4-Biphenylcarboxylic Acid

First add 4mL of 1mol·L^{-1} HCl and heat the stirred solution to about 70℃. Slowly add enough ethanol to the solution so that all material dissolves (approximately 30mL EtOH). Let the solution cool to room temperature and then place the beaker into an ice bath for 15min to complete the recrystallization. Isolate the crystalline product by vacuum filtration. A few milliliters of cold ethanol can be used to wash product out of the beaker if necessary. Leave the crystals to dry on the filter for at least 5 minutes (the structure of 4-biphenylcarboxylic acid is shown in Fig.3.2).

HOOC—

Fig.3.2 Chemical structure of 4-biphenylcarboxylic acid

Words and Expressions

purification *n.* 纯化
suction filtration 抽滤
petroleum ether 石油醚
chloroform *n.* 氯仿(三氯甲烷)
acetone *n.* 丙酮
ethylacetate *n.* 乙酸乙酯
ethanol *n.* 乙醇
impurity *n.* 杂质
reflux condenser 回流冷凝管
boiling chip 沸石
charcoal *n.* 活性炭
celite *n.* 硅藻土
mother liquor 母液
vacuum oven 真空干燥箱
drying pistol 干燥枪
filter flask 吸滤瓶
vacuum line 真空管路
aluminium foil 铝箔纸

Notes

➢ Place the crude compound (always keep a few “seed” crystals) in a flask fitted with a reflux condenser, add boiling chips and a small portion of solvent, and heat in a water bath.

参考译文：将粗品（始终保留一些“种子”晶体）放入装有回流冷凝器的烧瓶中，加入沸石和少量溶剂，并在水浴中加热。

➢ If you are using a mixed solvent, dissolve the crude in a small volume of the good solvent, heat to reflux, add the poor solvent in portions until the compound just begins to precipitate (cloudiness), add a few drops of the good solvent to redissolve the compound, and allow to cool.

参考译文：如果使用混合溶剂，则将粗品用少量良溶剂溶解，加热至回流，然后分批加入不良溶剂，直到刚刚开始沉淀（混浊），再加几滴良溶剂将沉淀的组分重新溶解，然后再冷却。

➢ As soon as all of the mother liquor has drained through the funnel, remove the suction and pour some cold solvent over the crystals, stir them if necessary, in order to ensure that they are thoroughly washed.

参考译文：尽快将全部母液通过漏斗抽滤，移除吸管并将一些冷溶剂倒在晶体上洗涤，必要时进行搅拌，以确保它们被清洗彻底。

➢ If they are placed in a dish, a beaker, or a sample vial, cover with aluminium foil, secure with wire or an elastic band, and punch a few small holes in the foil.

参考译文：如果晶体被放置在盘子、烧杯或者样品瓶中，需用铝箔纸包住，并用线或者橡皮筋绑住固定，然后在铝箔纸上扎一些小洞。

Exercises

1. Discuss the following questions

(1) What is the basic principle of recrystallization?

(2) Try to describe the process of recrystallization.

2. Translate the following into Chinese

(1) Purification of a solid by recrystallization from a solvent depends upon the fact that different substances are soluble to different extents in various solvents.

(2) In this case, the sample is dissolved in just enough of the hot solvent to form a saturated solution, the solution is cooled, and the crystals, which will be separated upon cooling, are collected by suction filtration.

(3) The most commonly used solvents in order of increasing polarity are petroleum ether, toluene, chloroform, acetone, ethyl acetate, ethanol, and water.

(4) In many cases a mixed solvent must be used, and combinations of toluene, chloroform, or ethyl acetate, with the petroleum ether fraction of similar boiling point are particularly useful.

(5) The difficulty here is that the compound tends to crystallize during the filtration so an excess of solvent (ca. 5%) should be added, and the apparatus used for the filtration should be preheated to about the boiling point of the solvent.

3. Put the following into English

抽滤　　杂质　　沸石　　母液

4. Translate the following into English

(1) 氯仿和二氯甲烷很少单独使用，因为它们是绝大多数有机化合物的良溶剂。

(2) 在添加溶剂时，如果粗品被类似硅胶、硫酸镁的不溶杂质污染，那么易导致溶剂被加过量。

(3) 如果溶液颜色很深并且（或者）含有少量焦油杂质，则需冷却后，加入约 2%（质量分数）的活性炭，回流几分钟，滤去活性炭。

(4) 如果使用沸点较高的溶剂（例如甲苯）进行重结晶，则必须在真空下加热样品数小时以确保去除所有溶剂。

UNIT 3 Distillation

Distillation is the most important and widely used method for the purification of organic liquids and the separation of liquid mixtures. The procedure involves boiling the liquid to vaporize it, and then condensing the vapor to give the distillate. The separation of a pair of liquids whose boiling points differ by 50~70℃ or more can be carried out by simple distillation, but if the difference is less, more complicated apparatus is required and the process is known as fractional distillation. Some liquids have boiling points that are too high to allow distillation at atmospheric pressure without causing thermal decomposition. Reducing the pressure lowers the boiling point and thus allows very high-boiling liquids and oils to be distilled easily and safely. The technique is known as vacuum distillation.

1. Simple Distillation

Distillation occurs when a liquid substance is heated and its vapors are allowed to condense in a vessel different from that used for heating. When a pure substance is distilled, a simple distillation is affected. What actually happens in this process is that the liquid is heated in a vessel until it vaporizes. The vapor passes into a condenser and is reconverted to the liquid, which is then collected in a receiver flask (Fig.3.3).

A system of this type is not very efficient and will not give a clean separation of liquids with a boiling point (b.p.) difference less than 50~70℃. If it is used for a mixture where the boiling points are closer, then, although the more volatile component will distill over first, it will be contaminated with the higher-boiling component even in the early stages of the distillation.

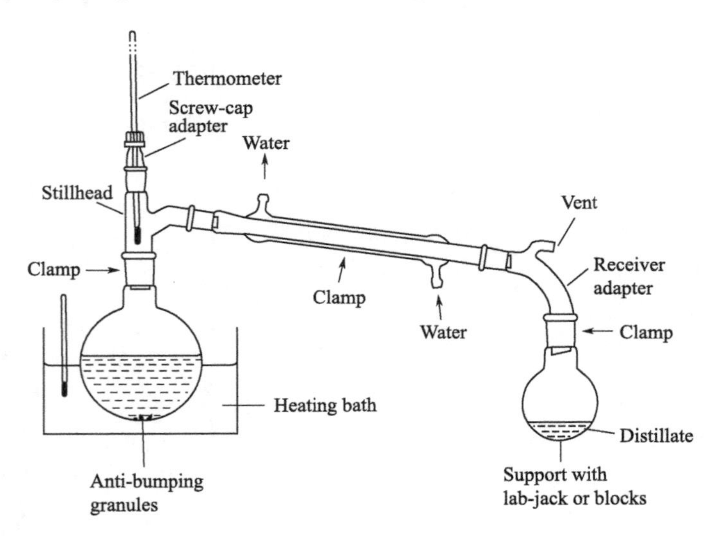

Fig.3.3 Simple distillation apparatus

2. Fractional Distillation

Fractional distillation is the most common of all distillation procedures (Fig.3.4). The efficiency of separation can be improved markedly by the use of a fractionating column. At best this will allow the separation of liquids whose boiling points differ by only a few degrees Celsius. The fractionating column consists of a vertical tube mounted above the distillation flask, containing packing material of high surface area onto which partial condensation of the vapor takes place. The objective is to achieve the most intimate contact between the ascending vapor and the descending condensate, which flows continuously down the column and back into the distillation pot. Under these conditions, the column will reach a state of equilibrium in which the vapor emerging at the top consists ideally of the more volatile component only. This process is known as contact rectification and its efficiency depends on maximizing the surface area of the descending liquid within the fractionating column.

Thermometer
Condenser
Vigreux column
Vacuum jacket
Air (N_2) leak (for vacuum distillation)

(a) Vigreux column

Condenser
Thermometer
Stillhead
Heating jacket
Receiver flask
Column packing
Fractionating column
Packing support

(b) Packed column

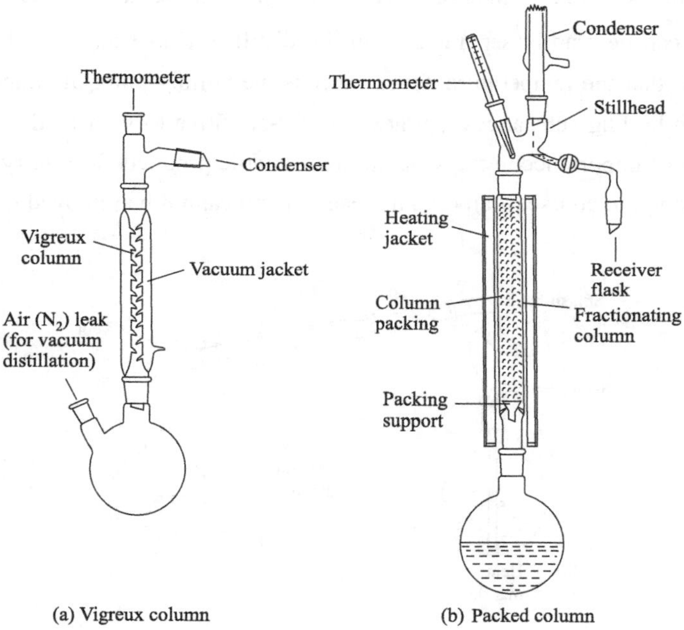

Fig.3.4 Fractional distillation

3. Vacuum Distillation

Many organic compounds cannot be distilled satisfactorily under atmospheric pressure because they undergo partial or complete thermal decomposition at their normal boiling points. Reducing the pressure to less than 30mmHg (1mmHg = 133.3Pa) considerably lowers the boiling point and this will usually allow distillation to be carried out without danger of decomposition.

The reduction in the boiling point will depend on the reduction in pressure and it can be estimated from a pressure-temperature nomograph. To find the approximate boiling point at any pressure, simply place a ruler on the central line at the atmospheric boiling point of the compound, pivot it to line up with the appropriate pressure marking on the right-hand line, and read off the predicted boiling point from the left-hand line. You can also use the nomograph to find the b.p. at any pressure if you know the b.p. at some other pressure, by first using the known data to arrive at an estimate of the atmospheric boiling point.

Two kinds of vacuum pumps are in common use: ①The water pump (sometimes known as a filter pump or aspirator) will reduce the pressure to 10~20mmHg. This will lower boiling points by 100~125℃. ②The rotary oil pump will give pressures as low as 0.01mmHg, below 30mmHg the b.p. is lowered by 10℃ each time the pressure is halved.

4. Steam Distillation

Some organic compounds that are virtually immiscible with water may be separated from non-volatile impurities including inorganic contaminants by steam distillation. This process is essentially a co-distillation with water and is usually accomplished by passing a current of steam through a hot mixture of the material to be distilled and water. Provided that the compound possesses an appreciable vapor pressure (5mmHg or more at 100℃), it will be carried over with the steam and, being immiscible, can be readily separated from the distillate (Fig.3.5). One of the advantages of steam distillation is that the temperature never exceeds the boiling point of water. This permits the purification of high-boiling substances that are too heat-sensitive to withstand ordinary distillation. The method is also of importance in the separation of volatile products from tarry material, which is often produced during the course of an organic reaction and cannot be removed easily by distillation or crystallization.

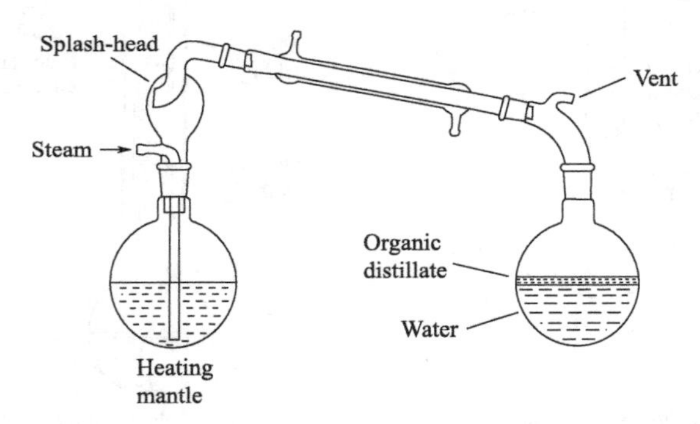

Fig.3.5 Steam distillation

5. Small-scale Distillation

The purification of small amounts of liquid (1~10mL) by distillation is difficult and can involve severe loss of material unless the apparatus is chosen with care. The major problem is the high percentage loss of material caused by hold-up, i.e., the unrecoverable material that forms a film over the surface of the flask, condenser and other glassware. This can be minimized by using very small apparatus designed to have a minimum wetted area. All small distillation flasks are pear-shaped to minimize thermal decomposition.

Amounts in the 1~10mL range can be distilled using apparatus of the type shown in Fig.3.6 in which the cold-finger condenser is integrated into the receiver adapter. For amounts at the top end of the range, the flask can be of the type shown in Fig.3.6(a), with a short Vigreux column to give some fractionation, and fitted with a capillary leak. However, for smaller amounts, the simple pear-shaped flask [Fig.3.6(b)] is preferable. It is impractical to use a capillary leak in these very small flasks and bumping is prevented by loosely packing the lower part of the bulb with glass wool, or by using pumice powder as an anti-bumping agent. Flask sizes down to 2mL can be used.

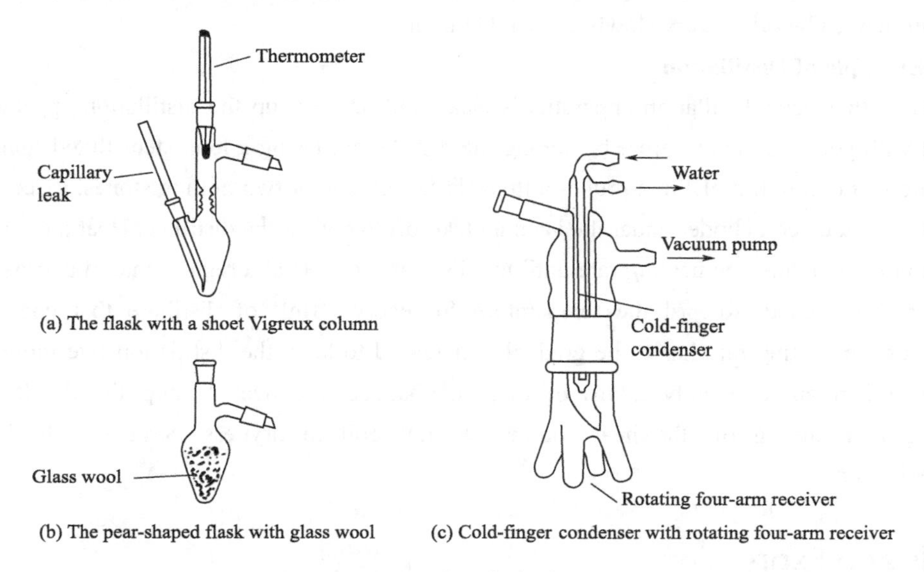

(a) The flask with a shoet Vigreux column

(b) The pear-shaped flask with glass wool

(c) Cold-finger condenser with rotating four-arm receiver

Fig.3.6 Small-scale distillation

6. Molecular Distillation

Normal vacuum distillation at pressures down to 10^{-1}mmHg is not practicable for some liquids either because they have very high boiling points or because of thermal instability. In many cases, such materials can be purified at very low pressure (10^{-6}~10^{-3}mmHg) by molecular distillation in a still designed, so that the gap between the liquid surface and the condenser is less than the mean free path of the molecules. The simplest version of such a still is shown in Fig.3.7. In molecular

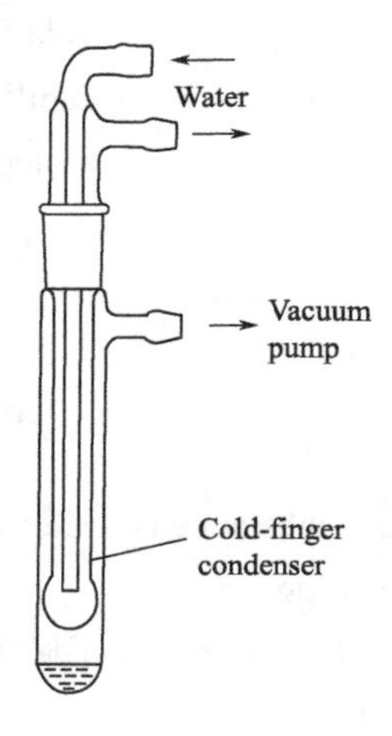

Fig.3.7 Molecular distillation

distillation, the molecules proceed directly from the liquid surface to the condenser without having to pass through a barrier of air molecules and few therefore return to the liquid. The normal concept of "boiling point" does not apply since there is no longer an equilibrium between liquid and vapor.

The high vacuum required can be produced by a diffusion pump backed by a rotary oil pump. In small-scale distillations, the vacuum must be applied very gently to avoid frothing and the bath temperature must be raised very slowly to avoid bumping.

7. Example of Distillation

Ensure that your distillation apparatus is clean and dry. Set up the distillation apparatus and check it with your instructor before beginning the distillation. Using a long stem fluted funnel, add the unknown mixture (22mL) to a round-bottom flask with one or two boiling stones. Place a clean, dry 100mL graduated cylinder under the vacuum take-off to collect the distillate. Heat the mixture to a gentle boil and adjust the heating rate until the distillate collects at a regular rate of approximately one drop per second. Record the temperature for every ~1mL of distillate that comes over. Sometimes the heating rate has to be gradually increased to keep the distillation rate more or less uniform. When approximately 20mL of distillate has been collected, stop the distillation by removing and turning off the heat source. Do not boil to dryness. Save the distillate for characterization.

Words and Expressions

distillation *n.* 蒸馏；蒸馏法
distillate *n.* 馏出物
boiling point 沸点
vaporize *v.* 蒸发
stillhead *n.* 分馏头
condenser *n.* 冷凝管
receiver *n.* 接收器
fractional distillation 分馏
distillation column 蒸馏柱
Vigreux column 维格勒柱
packed column 填充柱
splash head 防溅头
vacuum distillation 真空蒸馏
pump *n.* 泵
vent *n.* 出气口
steam distillation 水蒸气蒸馏
volatile *adj.* 易挥发的
capillary *n.* 毛细管
molecular distillation 分子蒸馏

Notes

➢ Distillation occurs when a liquid substance is heated and its vapors are allowed to condense in a vessel different from that used for heating.

参考译文：当加热一个液体物质，并使其蒸汽在不同于加热所用的容器中冷凝时就发生了蒸馏。

➢ The objective is to achieve the most intimate contact between the ascending vapor and the descending condensate, which flows continuously down the column and back into the distillation pot.

参考译文：其目的是在上升的蒸汽和下降的冷凝液之间实现最紧密的接触，而该冷凝液连续不断地从塔中向下流并回到蒸馏罐中。

➢ The method is also of importance in the separation of volatile products from tarry material, which is often produced during the course of an organic reaction and cannot be removed easily by distillation or crystallization.

参考译文：该方法对于从焦油状物质中分离挥发性产物也很重要，而焦油状物质常常在有机反应中产生并且难以通过蒸馏或结晶方法除去。

➢ The major problem is the high percentage loss of material caused by hold-up, i.e., the unrecoverable material that forms a film over the surface of the flask, condenser and other glassware.

参考译文：主要问题是滞留引起的物质高损失率，即不可回收的材料在烧瓶、冷凝器和其他玻璃器皿表面形成一层薄膜。

➢ In many cases, such materials can be purified at very low pressure (10^{-6}~10^{-3} mmHg) by molecular distillation in a still designed, so that the gap between the liquid surface and the condenser is less than the mean free path of the molecules.

参考译文：在许多情况下，此类材料可以在非常低的压力（10^{-6}~10^{-3}mmHg）下，通过蒸馏器中的分子蒸馏进行纯化，这样液体表面和冷凝器之间的间隙小于分子的平均自由程。

Exercises

1. Discuss the following questions

(1) Please summarise the principle of distillation.

(2) How the fractional distillation improves the efficiency of separation?

2. Put the following into Chinese

fractionating column　　fractional distillation　　steam distillation

receiver　　condenser　　thermometer　　Vigreux column　　capillary

3. Put the following into English

真空蒸馏　　蒸气压　　易挥发的　　混合物　　填充柱　　摄氏温度

4. Translate the following into Chinese

(1) Distillation is the most important and widely used method for the purification of organic liquids and the separation of liquid mixtures.

(2) A system of this type is not very efficient and will not give a clean separation of liquids with a boiling point (b.p.) difference of less than 50~70℃.

(3) The fractionating column consists of a vertical tube mounted above the distillation flask, containing packing material of high surface area onto which partial condensation of the vapor takes place.

(4) Many organic compounds cannot be distilled satisfactorily under atmospheric pressure

because they undergo partial or complete thermal decomposition at their normal boiling points.

(5) This permits the purification of high-boiling substances that are too heat-sensitive to withstand ordinary distillation.

(6) In small-scale distillations, the vacuum must be applied very gently to avoid frothing and the bath temperature raised very slowly to avoid bumping.

(7) The fundamental separation process in refining petroleum is fractional distillation. Practically, all crude petroleum that enters a refinery goes to distillation units, where it is heated to temperatures as high as 370℃ to 425℃ and separated into fractions. Each fraction contains a mixture of hydrocarbons that boils within a particular range.

UNIT 4 Separation

Extraction and chromatography are the most universally important techniques in organic chemistry. Extraction takes the advantage of the different solubility of compounds in organic solution and aqueous solution to separate them. However, not all the compounds possess large solubility difference are separated by extraction. Compared to extraction, chromatography utilizes the different adsorption affinity of compounds to adsorbent and ability to be eluted by different organic solvents which possess higher separation ability.

1. Extraction Procedures

The most straightforward extractions are carried out in a separating funnel (Fig.3.8). The procedure involves mixing an organic solution (in a solvent that is not miscible with water) with water (or aqueous acid or alkali), shaking the funnel to mix the two layers thoroughly, allowing the mixture to stand until the two layers have separated again, and then running off the lower layer into another vessel. The mixing allows material to pass from one layer to the other depending on its relative solubility in each.

Basically, extractions are of two kinds: ①extraction with water to remove water-soluble material (usually inorganic) from a mixture, and ②extraction with aqueous acid or aqueous base to remove, respectively, organic bases or organic acids from the organic layer.

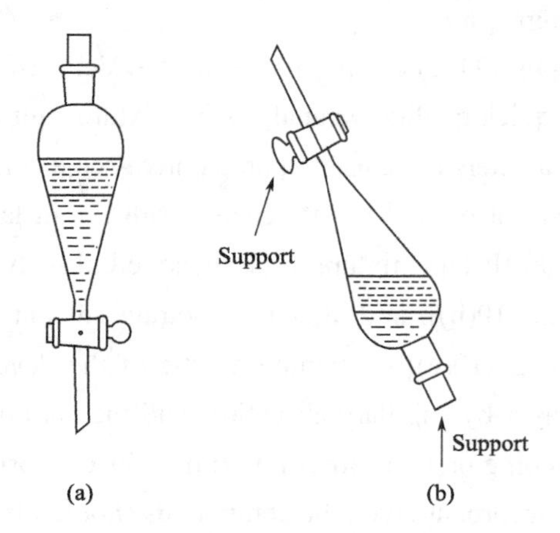

Fig.3.8 (a) Separating funnel; (b) Position for venting excess pressure

It is sometimes necessary to extract a solid organic product from a solid mixture containing other materials (usually inorganic) that are soluble in neither organic solvents nor water. This is easier to do if the mixture is first grounded up finely with a mortar and pestle. If the organic solid is readily soluble in an organic solvent, then the separation is easily achieved, e.g., by putting the mixture into a sintered funnel, adding the solvent, stirring and then applying suction. However, if the organic solid is not easily soluble then a continuous extraction method, using a Soxhlet extractor,

must be used. The mixture is placed in the porous thimble. In operation, the solvent distills into the extraction chamber and, when full, siphons back into the flask. Prolonged repetition of this cycle slowly transfers the solid into the flask (Fig.3.9).

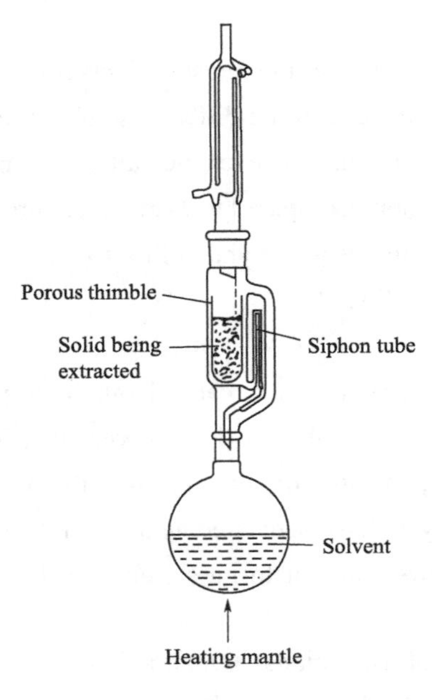

Fig.3.9 Soxhlet continuous extractor for the extraction of solids

2. Thin-layer Chromatography

Thin-layer chromatography (TLC) is one of the most widely used forms of chromatography, and is of enormous value for quick qualitative analysis of mixtures, for monitoring reactions and for determining the operating parameters to be used in preparative-scale column chromatography.

The separation is carried out on a flat plate coated with a thin layer of an adsorbent such as silica gel or alumina [Fig.3.10(a)]. The mixture to be separated, dissolved in an appropriate solvent, is spotted onto the plate [Fig.3.10(b)], and after the spotting solvent has evaporated, the plate is placed in a developing jar [Fig.3.10(c)] containing a little of the developing solvent. The solvent rises through the adsorbent layer by capillary attraction, and the various compounds in the mixture ascend at different rates depending on their differing affinity for the absorbent. When the solvent has almost reached the top of the adsorbent layer, the compounds should ideally be well separated.

3. Column Chromatography

Column chromatography is the organic chemists' single most important technique for the separation of mixtures on a preparative scale (a few milligrams up to tens of grams). The separation process used is usually liquid-solid (adsorption) chromatography, since it works well for most non-ionic types of compounds.

The separation is carried out using a column of the adsorbent packed into a glass tube [Fig.3.11(a)] as a porous bed through which the mobile phase can flow. The mobile phase, generally known as the eluting solvent or eluent, is an organic solvent such as hexane. The mixture to be

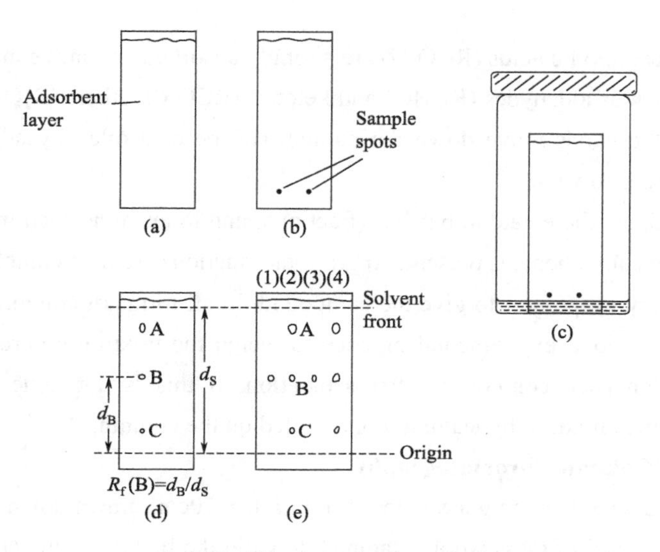

Fig.3.10 Thin-layer chromatography

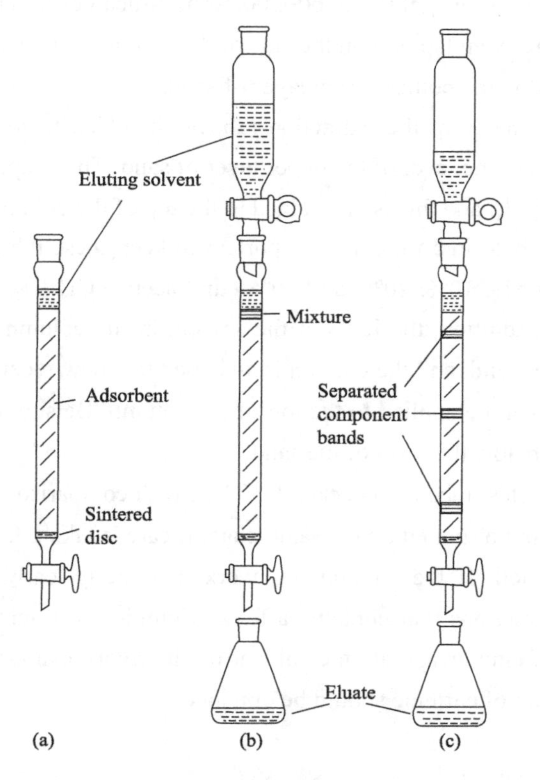

Fig.3.11 Column chromatography (with gravity elution)

separated is applied to the top of the column where it is adsorbed by the stationary phase. The eluent then passes continuously through the column [Fig.3.11(b)]. Each component in the mixture is carried down the column by the mobile phase at a speed that depends on its affinity for the adsorbent. Ideally the mixture will separate into a number of discrete bands [Fig.3.11(c)], which pass slowly down the column and eventually into a collecting vessel. Polar compounds such as alcohols (ROH), amines (RNH_2), or carboxylic acids (RCO_2H) are strongly adsorbed and move more slowly than less polar compounds such as aldehydes (RCHO) and ketones (RCOR'), ethers (R_2O) and hydrocarbons. The rate at which the bands move down the column can be controlled by adjusting the strength (polarity) of the eluting solvent.

It is usual to collect the eluate in batches (fractions) and to examine each by TLC to determine which (if any) of the components is present. Appropriate fractions are then combined and the solvent is removed on a rotary evaporator to give the compound. In all forms of column chromatography, it is an essential part of good experimental practice to weigh the mixture before putting it onto the column and to weigh each component after separation. If this is not done, you may "lose" a compound in a complex mixture by leaving it undetected on the column.

4. Example of Column Chromatography

Commercially available open glass tubes, 19mm i.d. × 10cm, drawn down at one end to a drip tip, are used. A small wad of glass wool is tamped down in the bottom of the column, then covered with a 0.5cm layer of sand. Silica gel (0.7g, 60~200mesh, "silica gel for chromatography") is added in a thin stream, with concurrent tapping on the side of the column to free air bubbles. After the silica gel is settled, it is covered with another 0.5cm layer of sand.

The residue of caffeine from the extraction (above) is added to the top of the column. The solvent is allowed to drop to the level of the upper layer of sand. The evaporating flask is rinsed with an additional 5mL of CH_2Cl_2, and this is also added to the top of the column. Eluate from the column is collected in test tubes, held in a rack. It is important to keep these tubes in order. The column is eluted with 5mL each of 5%, 10%, 20%, and 40% ethyl acetate/CH_2Cl_2, followed by 5mL of pure ethyl acetate. After each addition, the level of the solvent in the column is allowed to drop to the level of the upper layer of sand, and the column is switched to a new receiving tube. If elution is too slow, gentle air pressure can be applied to the top of the column. Be sure to remove the air pressure before the solvent drops below the level of the sand.

The contents of each test tube are checked by TLC, with comparison to an authentic sample of caffeine. Several tubes can be spotted on the same plate, if care is taken. It helps to mark the spotting points lightly with a pencil. If the column has worked properly, early yellow fractions will be followed by colourless fractions that contain caffeine. Combine in a tared round-bottom flask the fractions that contain caffeine, evaporate the solvent on the rotary evaporator, and weigh the fluffy white residue. About 65mg of caffeine should be obtained.

Words and Expressions

extraction *n.* 萃取

miscible *adj.* 易混溶的

separating funnel 分液漏斗

Soxhlet extractor 索氏提取器

mortar *n.* 研钵

pestle *n.* 研杵

sintered funnel 烧结漏斗

extraction chamber 萃取室

siphon tube　虹吸管
thin-layer chromatography　薄层色谱
column chromatography　柱色谱
silica gel　硅胶
mobile phase　流动相
stationary phase　固定相
eluent　*n.* 洗脱剂
alumina　*n.* 氧化铝
developing jar　展开缸
capillary attraction　毛细吸引
porous bed　多孔床
liquid-solid chromatography　液固色谱

Notes

➢ Thin-layer chromatography (TLC) is one of the most widely used forms of chromatography, and is of enormous value for quick qualitative analysis of mixtures, for monitoring reactions and for determining the operating parameters to be used in preparative-scale column chromatography.

参考译文：薄层色谱是使用最为广泛的色谱方法之一，对于快速定性分析混合物、监测反应和确定柱色谱制备使用的操作参数具有巨大的价值。

➢ The solvent rises through the adsorbent layer by capillary attraction, and the various compounds in the mixture ascend at different rates depending on their differing affinity for the adsorbent.

参考译文：溶剂通过毛细吸引从吸附层上升，混合物中各种化合物根据其对吸附剂的不同亲和力以不同的速率上升。

➢ The separation is carried out using a column of the adsorbent packed into a glass tube as a porous bed through which the mobile phase can flow.

参考译文：将吸附剂填充到玻璃管中作为多孔床进行分离，流动相可以流过该多孔床。

➢ Each component in the mixture is carried down the column by the mobile phase at a speed that depends on its affinity for the adsorbent.

参考译文：混合物中的每种组分都通过流动相以一定的速度向下流动，该速度取决于其对吸附剂的亲和力。

➢ It is usual to collect the eluate in batches (fractions) and to examine each by TLC to determine which (if any) of the components is present.

参考译文：通常分批（部分）收集洗脱液并通过 TLC 检查每个洗脱液以确定存在哪种（如果有）成分。

Exercises

1. Discuss the following questions

(1) How to extract organic compounds from aqueous solutions?

(2) What is the separation principle of column chromatography?

2. Put the following into Chinese

funnel　　extraction chamber　　adsorbent　　developing solvent

mobile phase　　stationary phase　　eluent

3. Put the following into English

展开缸　　硅胶　　液固色谱　　毛细吸引　　多孔床

4. Translate the following into Chinese

(1) The mixing allows material to pass from one layer to the other depending on its relative solubility in each.

(2) The separation is carried out on a flat plate coated with a thin layer of an adsorbent such as silica gel or alumina.

(3) The rate at which the bands move down the column can be controlled by adjusting the strength (polarity) of the eluting solvent.

(4) The contents of each test tube are checked by TLC, with comparison to an authentic sample of caffeine.

(5) Both procedures (qualitative TLC of analgesics, column chromatography) can be carried out within a single 3h laboratory period, if the TLC determinations are done at slow periods of the extraction (e.g., while the hot aqueous extract is cooling). We currently take two laboratory periods and include quantitative determination of the same "unknown" analgesic mixture, using high performance liquid chromatography.

UNIT 5 Scale-up Synthesis

1. Introduction

The Suzuki-Miyaura cross-coupling reaction is undeniably one of the most powerful reactions in modern organic synthesis for the construction of carbon-carbon bonds. The Suzuki-Miyaura cross-coupling reaction has been extensively used to access functionalized styrene derivatives in large scale production of active pharmaceutical ingredients via the reaction of nucleophilic vinylboron species with various aryl electrophiles, including aryl halides, pseudo-halides, and diazonium salts.

LSZ102 is a clinical development candidate currently in phase I/Ib trials for the treatment of ERα positive breast cancer. The synthesis of such a molecule can be challenging and needs to respond to supply needs, quality requirements, and efficiency constraints. During early stage development, the quality of the drug substance (DS) LSZ102 was mainly driven by the outcome of the Suzuki-Miyaura cross-coupling and in particular by the impurities generated.

2. Challenge

Initially, the Suzuki coupling of **1** (1eq.) and **2** (1.7eq.) uses tetrakis (triphenylphosphine) palladium $Pd(PPh_3)_4$ (7.5%, mole fraction) in a mixture of 1,4-dioxane and water (Fig.3.12).

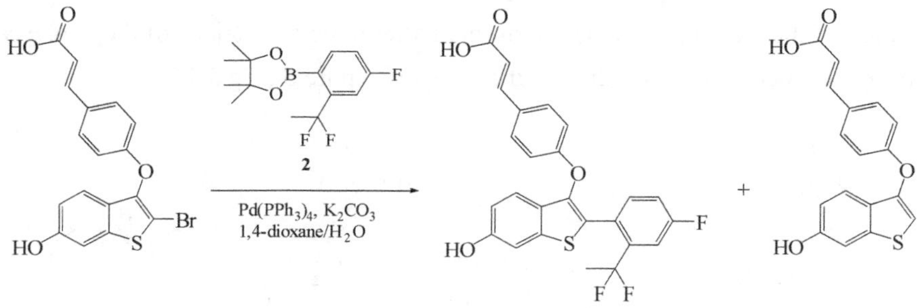

Fig.3.12 Small scale Suzuki reaction scheme

The main impurity of the cross-coupling is the debromination product **4**. A large excess of the custom-manufactured boronate ester is required to ensure sufficient conversion, which results in a dramatic impact on the overall cost. Moreover, considering the high catalyst loading, a tedious series of treatment is required postreaction to achieve sufficient palladium removal and meet specifications. An extensive workup is also required to remove the apolar, nonacidic impurities. The overall protocol eventually turns quite unproductive and highly resource intensive.

3. Improved Method

Based on many encouraging applications of the surfactant technology onto such sensitive

Suzuki-Miyaura cross-coupling systems, mild reaction conditions in addition to the micellar effect could be beneficial and allow for a reduction of prominent impurities. Surfactant TPGS-750-M in water is used as an alternative medium for the transformation (Fig.3.13).

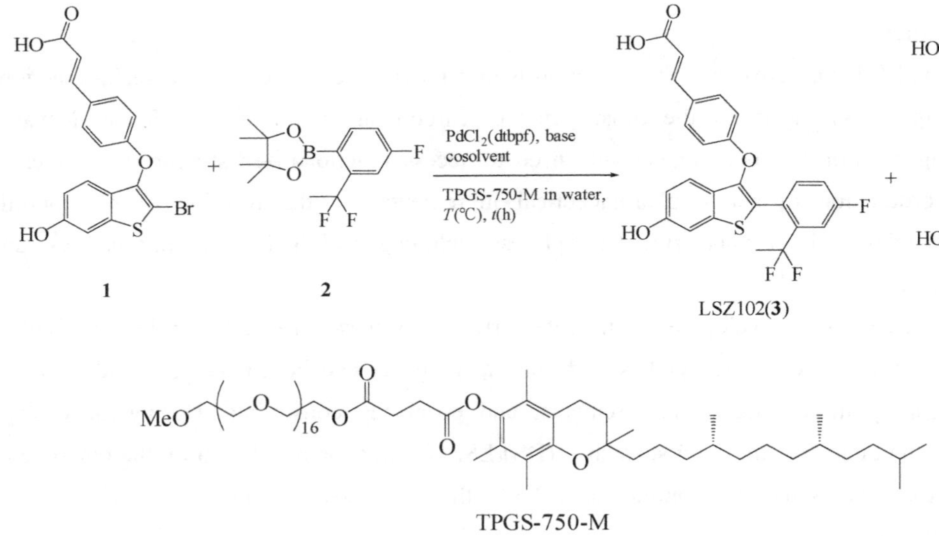

Fig.3.13 Typical Suzuki reaction scheme and structure of TPGS-750-M

The addition of dicyclohexylamine nicely leads to the crystalline dicyclohexylamine salt crude **5**. The liberation of the free acid **1** is then performed after three extractions of dicyclohexylamine by *n*-heptane from an aqueous basic solution of the compound pure **5** (Fig.3.14).

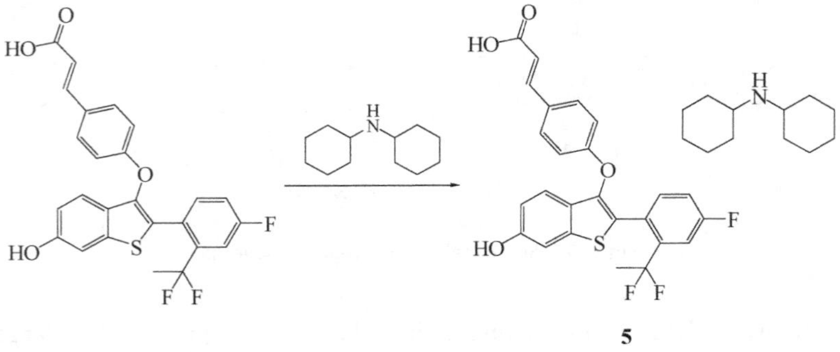

Fig.3.14 Salinization of LSZ102

4. Scale-up

Different reaction conditions are explored (Table 3.1). Adding the base as a concentrated solution and decreasing the catalyst loading to 1.5% (mole fraction) leads to full conversion of the desired cross-coupled compound **3** after 6.5h, and generating only a minimal amount of side product **4** (<0.1%).

Table 3.1 Screened reaction conditions

Entry	Cosolvent	T/℃	t/h	Base	Conv./%	4 concentration/%	Additive
1	1,4-dioxane/H_2O	70	14	K_2CO_3	>98	7~8	N/A
2	THF	25	12	Et_3N	75	4	N/A
3	THF	40	12	K_3PO_4	90	7	N/A
4	THF	40	12	K_3PO_4	86	9	N/A
5	PEG200	40	12	K_3PO_4	87	9	N/A
6	Acetone	40	12	K_3PO_4	97	1	N/A
7	Acetone	40	3	K_3PO_4	96	3	N/A
8	Acetone	40	3	K_3PO_4	>99	1	N/A
9	Acetone	40	6.5	K_3PO_4	>99	<0.1	LiBr

With this set of conditions in hand, a scale-up reaction can be conducted in scale-up facilities. In TPGS-750-M (20V) are added the starting materials **1** (1.0eq.) and **2** (1.3eq.) at room temperature, followed by LiBr (1.0eq.) and the palladium catalyst (1.5%), leading to a beige suspension. Acetone (2V) is added, and the mixture is warmed to 40℃ and stirred for 15min. A solution of K_3PO_4 in water (5mol·L^{-1}) is then added over 30min. The resulting mixture is stirred for 7h at 40℃.

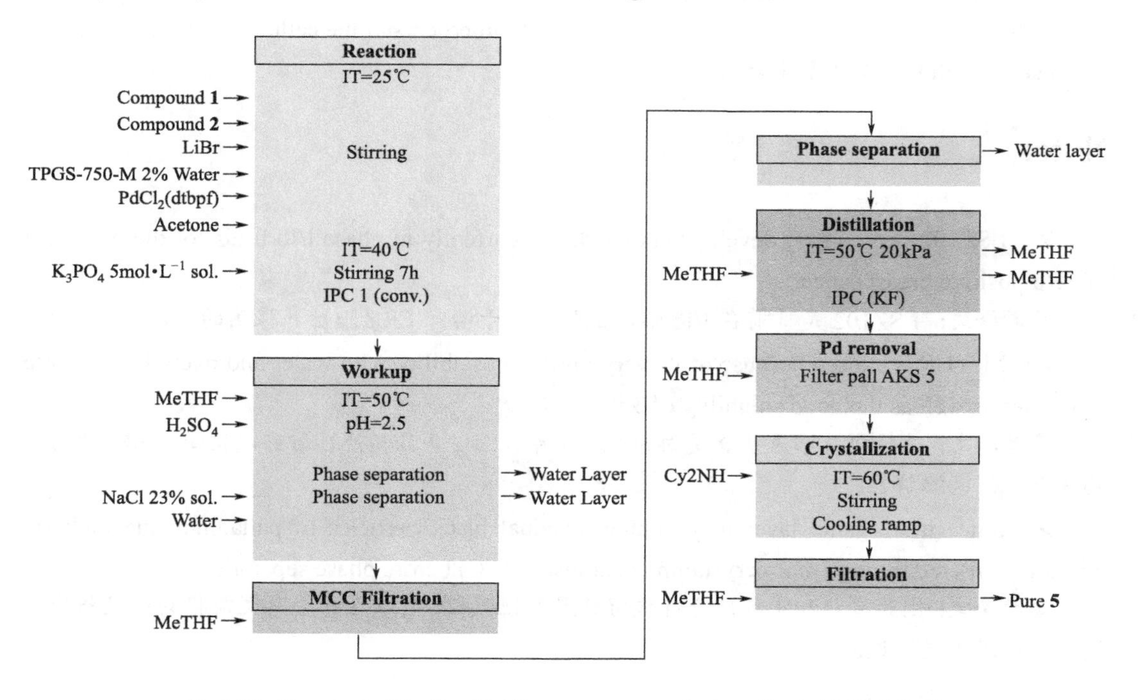

Fig.3.15 Flowchart of the Suzuki-Miyaura cross-coupling reaction in TPGS-750-M

The salt forming step using dicyclohexylamine as base can isolate the compound pure **5** and then release the free acid **3** by sequential pH switch and extraction of organic substances using an organic solvent. MeTHF is chosen because of its very limited miscibility with water and overall acceptable properties as well as the good solubility of **3** in it. Because of the use of potassium phosphate in excess, the pH value of the solution at the end of the reaction is typically 9.3~9.6 and needs first to be adjusted to an acidic pH value to release the free acid **3** in the organic layer. Addition of conc. H_2SO_4 to reach pH 2.5 at 50℃ allows for a good phase separation. The organic layer is washed successively with a NaCl solution (23%, mass fraction) and with pure water. Since the organic layer may contain residual black particles of palladium, the biphasic mixture is filtered through microcrystalline cellulose (MCC) before phase separation. Azeotropic distillation ensures a low level of water content before the salt forming step. Dicyclohexylamine is added at 60℃, followed by seed crystals. The crystallization is induced upon cooling. The desired product pure **5** is obtained by filtration followed by a drying step (Fig.3.15).

Words and Expressions

styrene derivative 苯乙烯衍生物
vinylboron species 乙烯硼化物
boronate ester 硼酸酯
active pharmaceutical ingredient 活性药物原料
pseudo-halide 类卤素化合物
surfactant *n.* 表面活性剂
medium *n.* 介质
cosolvent *n.* 助溶剂
miscibility *n.* 混溶性
microcrystalline cellulose 微晶纤维素

Notes

➢ LSZ102 is a clinical development candidate currently in phase I/Ib trials for the treatment of ERα positive breast cancer.

参考译文：LSZ102 是目前在 I/Ib 临床试验中用于治疗 ERα 阳性乳腺癌的候选药物。

➢ MeTHF is chosen because of its very limited miscibility with water and overall acceptable properties as well as the good solubility of **3** in it.

参考译文：选择甲基四氢呋喃是因为其与水难互溶，并且其他性质可以接受，同时对化合物 **3** 有很好的溶解性。

➢ Since the organic layer may contain residual black particles of palladium, the biphasic mixture is filtered through microcrystalline cellulose (MCC) before phase separation.

参考译文：因为有机相中可能含有残留的黑色钯颗粒，所以在相分离之前将两相混合物通过微晶纤维素进行过滤。

Exercises

1. Discuss the following question

What is the typical process to scale-up a reaction?

2. Translate the following into Chinese

(1) The Suzuki-Miyaura cross-coupling reaction is undeniably one of the most powerful reactions in modern organic synthesis for the construction of carbon-carbon bonds.

(2) With this set of conditions in hand, a scale-up reaction can be conducted in scale-up facilities.

(3) The salt forming step using dicyclohexylamine as base can isolate the compound pure **5** and then release the free acid **3** by sequential pH switch and extraction of organic substances using an organic solvent.

3. Put the following into English

硼酸酯　　表面活性剂　　介质　　助溶剂　　混溶性

4. Translate the following into Chinese

(1) 表面活性剂技术在敏感的 Suzuki-Miyaura 交叉偶联体系中有许多令人兴奋的应用，除了胶束效应外，温和的反应条件可能是有益的，可以减少主要的杂质。

(2) 在水中表面活性剂 TPGS-750-M 被用作转化的替代介质。

PART 4 Fine Chemicals

UNIT 1 General Introduction of Fine Chemicals

The underlying principle for definition of the term "fine chemicals" is a three-tier segmentation of the universe of chemicals into commodities, fine chemicals, and specialty chemicals (Fig.4.1).

Commodities	Fine chemicals	Specialty chemicals
Single pure chemical substances	Single pure chemical substances	Mixtures
Produced in dedicated plants	Produced in multi-purpose plants	Formulated
High volume/ low price	Low vol. (<1000 Mt/a) high price (> $10/kg)	Undifferentiated
Many applications	Few applications	Undifferentiated
Sold on specifications	Sold on specifications "what they are"	Sold on performance "what they can do"

Fig.4.1 The underlying principle of the term "fine chemicals"

Commodities are large-volume, low-price, homogeneous, and standardized chemicals produced in dedicated plants and used for a large variety of applications. Prices, typically less than $1/kg, are cyclic and are fully transparent. Petrochemicals, basic chemicals, heavy organic and inorganic chemicals, (large-volume) monomers, commodity fibres, and plastics are all part of commodities. Typical examples of single products are ethylene, propylene, acrylonitrile, caprolactam, methanol, toluene, *o*-xylene, phthalic anhydride, poly(vinylchloride) soda, and sulfuric acid.

Fine chemicals are complex, single, pure chemical substances. They are produced in limited quantities (up to 1000Mt per year) in multipurpose plants by multistep batch chemical or biotechnological processes. They are based on exacting specifications, are used as starting materials for specialty chemicals, particularly pharmaceuticals, biopharmaceuticals and agrochemicals, and are sold for more than $10/kg. In China, fine chemicals refer to the key components to 11 categories of chemical industry fields, including pesticides, dyestuffs, coatings (including paints and inks), pigments, reagents and high-purity chemicals, information chemicals (including photosensitive materials and magnetic recording materials), food and feed additives, adhesives, catalysts and

音频

auxiliaries, chemical drugs and chemicals for daily use.

The category is further subdivided on the basis of either the added value (building blocks, advanced intermediates, or active ingredients) or the type of business transaction (standard or exclusive products). As the term indicates, exclusive products are made exclusively by one manufacturer for one customer, which typically uses them for the manufacture of a patented specialty chemical, primarily a drug or agrochemical. Typical examples of single products are β-lactams, imidazoles, pyrazoles, triazoles, tetrazoles, pyridines, pyrimidines, and other *N*-heterocyclic compounds. A third way of differentiation is the regulatory status, which governs the manufacture. Active pharmaceutical ingredients (APIs) and advanced intermediates thereof have to be produced under current good manufacturing practice (cGMP) regulations. They are established by the (U.S.) Food and Drug Administration (FDA) in order to guarantee the highest possible safety of the drugs made thereof. All advanced intermediates and APIs destined for drugs and other specialty chemicals destined for human consumption on the U.S. market have to be produced according to cGMP rules, regardless of the location of the plant.

A precise distinction between commodities and fine chemicals is not feasible. In very broad terms, commodities are made by chemical engineers and fine chemicals by chemists. Both commodities and fine chemicals are identified according to specifications. Both are sold within the chemical industry, and customers know how to use them better than suppliers. In terms of volume, the dividing line comes at about 1000t/a; in terms of unit sale prices, this is set at about $10/kg. Both numbers are somewhat arbitrary and controversial. Many large chemical companies include larger-volume/lower-unit-price products, so they can claim to have a large fine chemical business (which is more appealing than commodities). The threshold numbers also cut sometimes right into otherwise consistent product groups. This is, for instance, the case for APIs, amino acids, and vitamins. In all three cases, the two largest-volume products, namely, acetyl salicylic acid and paracetamol, L-lysine and D, L-methionine, and ascorbic acid and niacin, respectively, are produced in quantities exceeding 10000t/a, and are sold at prices below the $10/kg level.

Specialty chemicals are formulations of chemicals containing one or more fine chemicals as active ingredients. They are identified according to performance properties. Customers are mostly trades outside the chemical industry and the public. Specialty chemicals are usually sold under brand names. Suppliers have to provide product information. The distinction between fine and specialty chemicals is net. The former is sold on the basis of "what they are"; the latter, on "what they can do". In the life science industry, the active ingredients of drugs, also known as APIs or drug substances (DS), are fine chemicals, the formulated drugs specialties, as known as drug products.

1. Fine Chemical Manufacture

Processes in fine chemical manufacture differ from processes for the manufacture of commodities in many respects:

① A significant proportion of the fine chemicals are complexes, multifunctional large molecules. These molecules are labile, unstable at elevated temperature, and sensitive towards changes in their environment (e.g., pH). Therefore, processes are needed with inherent protective measures (e.g., chemical or physical quenching) or a precise control system to operate exactly within the allowable range. Otherwise the yield of the desired product can drop to nearly zero.

② Fine chemicals are high-added-value products. In general, expensive raw materials are processed to obtain fine chemicals and therefore, the degree of their utilization is very important. With complex reaction pathways, selectivity is the key problem to make the process profitable. Selectivity is significant also because of difficulties in isolation and purification of the desired product from many side products, especially those with physical-chemical properties similar to those of the desired product. Furthermore, a low selectivity results in large streams of pollutants to be treated before they can be disposed of. Selectivity is even more an issue because in contrast with bulk chemicals production, where a limited single-pass conversion coupled with separation and recycling of unreacted raw materials is often applied, usually complete conversion is aimed at. Selectivity can be controlled by chemical factors such as chemical routes, solvents, catalysts and operation conditions, but it is also strongly dependent on engineering solutions. Catalysts are the key to increase the selectivity.

③ In the manufacture of fine chemicals many hazardous chemicals are used, such as highly flammable solvents, cyanides, phosgene, halogens, volatile amines, isocyanates and phosphorous compounds. The use of hazardous and toxic chemicals produces severe problems associated with safety and effluent disposal. Moreover, fine chemistry reactions are predominantly carried out in batch stirred-tank reactors characterized by a large inventory of dangerous chemicals, and a limited possibility to transfer the generated heat to the surroundings. Therefore, the risk of thermal runaways, explosions, and emission of pollutants to the surroundings is greater than in bulk production. That is why much attention must be paid to safety, health hazards, and waste disposal during development, scale-up, and operation of the process.

Fine chemicals are often manufactured in multistep conventional syntheses, which results in a high consumption of raw materials and, consequently, large amounts of by-products and wastes. On average, the consumption of raw materials in the bulk chemical business is about 1kg/kg of product. This figure in the dine chemistry is much greater, and can reach up to 100kg/kg for pharmaceuticals. The high raw materials-to-product ratio in the fine chemistry justifies extensive search for selective catalysts. Use of effective catalysts would result in a decrease of reactant consumption and waste production, and the simultaneous reduction of the number of steps in the synthesis.

④ One of the most important features of the fine chemical manufacture is the great variety of the products, with new products permanently emerging. Therefore, significant fluctuations in the demand exist for a variety of chemicals. If each product would be manufactured using a plant dedicated to the particular process, the investment and labor costs would be enormous. In combination with the ever-changing demand and given the fact that plants are usually run below their design capacity, this would make the manufacturing costs very high. Therefore, only larger volume fine chemicals or compounds obtained in a specific way or of extremely high purity are produced in dedicated plants.

Most of the fine chemicals, however, are manufactured in multipurpose or multiproduct plants

(MPPs). They consist of versatile equipment for reaction and utilities. By changing the connection between the units and careful cleaning of the equipment to be used in the next campaign, one can adapt the plant to the intended process. The investment and labor costs are significantly lower for MPPs than for dedicated plants, while the flexibility necessary to meet changing demands is provided. The need for versatility of equipment originates from the great number of the products in rather limited quantities to be manufactured in the plant every year. Such versatile equipment is suitable for process, although it is certainly not optimal. The most versatile reactors are stirred-tank reactors operated in batch or semibatch mode, so such reactors are used in multiproduct plants. Continuous plants with reactors of small volume are sometimes used despite the small capacity required. This is the case when the residence time of reactants in the reaction zone must be short or when too many hazardous compounds could accumulate in the reaction zone.

From the foregoing, it will be clear that in fine chemical process development, the strategy differs profoundly from that in the bulk chemical industry. The major steps are adaptation of procedures to constraints imposed by the existing facilities with some necessary equipment additions, or choice of appropriate equipment and determination of the procedures for a newly built plant, in such a way that procedures in both cases guarantee the profitable, competitive, and safe operation of a plant.

⑤ The accuracy of analytical methods has increased enormously in the past decades and this has enabled detection of even almost negligible traces of impurities. The consequence is that both regulations and specifications for intermediates and final fine chemicals have become stricter. Therefore, very pure compounds must often be produced with impurities at 10^{-6} or 10^{-9} level. The production of complex molecules in many cases results in mixture containing isomers, including optical isomers. The demand for enantiomeric materials is growing at the expense of their racemic counterparts, driven primarily by the pharmaceutical industry. Both stereoselective synthesis and effective, often non-conventional methods of racemate resolution, stimulate the development of new stereoselective catalysts, including biocatalysts.

The implications of the features of fine chemical manufacture mentioned above, are, however, not that obvious to all parties taking part in process development. On the one hand, chemical and process engineers are dedicated mainly to the engineering part in process development, often neglecting process chemistry, as this might be considered less important for a full-scale plant. On the other hand, synthetic chemists often finish their work with laboratory recipes neglecting needs of the procedures for process development.

2. Microfluidic Reaction System

Ever since Wöhler's laboratory synthesis of urea in 1828, the chemist's toolkit has predominantly consisted of macroscopic components fabricated from glass. Examples include round-bottomed flasks, test tubes, distillation columns, reflux condensers and retorts. Despite advances in experimental and mechanistic organic chemistry during the past century, it is noteworthy that the basic experimental techniques and associated equipment have remained largely unchanged. There are a number of reasons why traditional synthetic chemistry is performed in the aforementioned equipment, but is there any advantage to performe synthetic chemistry in volumes 5~9 orders of magnitude smaller than those associated with bench-top chemistry? The application of techniques cultivated in semiconductor industries have allowed the creation of a new instrumental platform able to efficiently manipulate, process and analyze molecular reactions on the micrometer to nanometer scale (Fig.4.2). Even at this early stage in the development of "microfluidic" reaction systems, it is clear that advantages engendered by miniaturization may affect molecular synthesis similarly to the way that the integrated circuit has defined the computer revolution over the past 50 years.

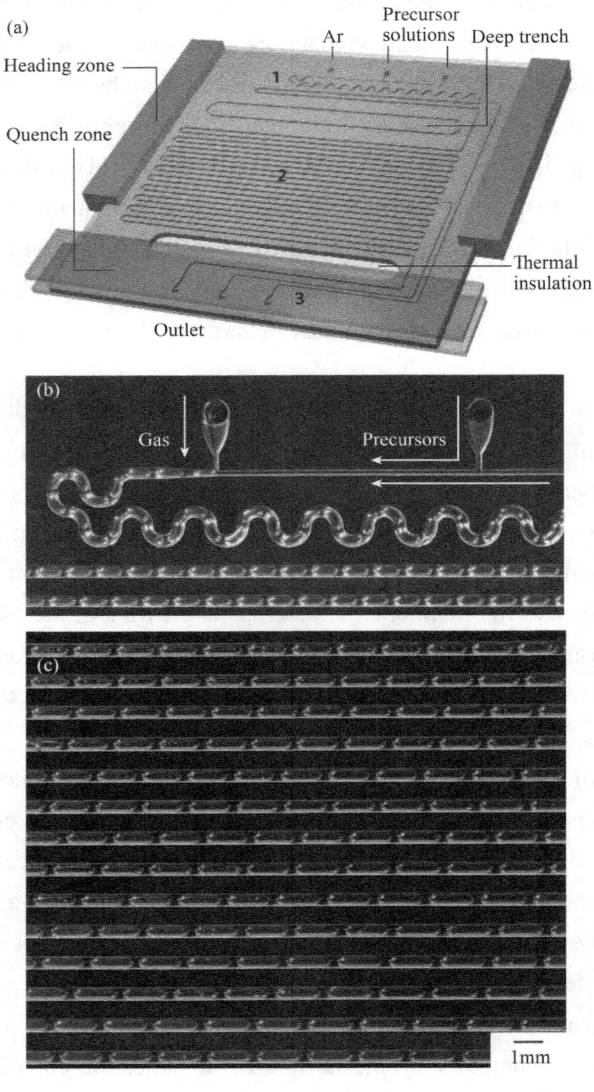

Fig.4.2　Microfluidic reactor for nanoparticle production

Notes: (a) The reactor allows rapid precursor mixing (sector 1), controlled particle growth (sector 2) and reaction quenching (sector 3). The reactor accommodates a ~1-metre-long reaction channel and two shallow side channels for collecting reaction aliquots, (b) photographs of heated inlets and (c) main channel section.

Developed in 1990s, microfluidics is the science and technology of systems that process or manipulate microfluidic fluids (in nanoliters to liters) through microchannels with a diameter from ten to hundreds of micrometers. In the last two decades, a variety of microfluidic devices and systems have been designed and optimized to be applied in diverse fields, such as single or multiple emulsions formation, continuous chemical synthesis, multifunctional microparticles preparation, bioinstrumentation, etc. For instance, because of its precise controllability and operating stability, microfluidic technology could be employed to prepare monodispersed droplets or micro-bubbles with polydispersity index less than 5%. Such a method plays a significant role in the fields of food science, cosmetic manufacture, and biomedical testing.

On the one hand, the heat and mass transfer coefficients within microchannels are proved to be one to two orders of magnitude higher than those of macroscale reactors and mixers, making uniform mixing, fast extraction, and reaction intensification possible in micromixers, microreactors, and micro-dispersers, respectively. On the other hand, because of the immiscibility and phase splitting nature of multiphase flow, particular attentions are paid on the application of microfluidic system with multiphase flow to intensify heterogeneous reactions, prepare multiphase emulsions, and fabricate anisotropic functional materials. In a word, with its great advances in last several decades, multiphase flows in microfluidic system have shown numerous advantages, like fastening heat and mass transfer, reducing energy consumption, improving productivity and purity, etc. Fortunately, because of the recent development and commercialization of micromachining technologies, microfluidic chips gradually moves from laboratory stage to practical application stage.

Words and Expressions

fine chemical 精细化学品
commodity chemical 通用化学品
specialty chemical 专用化学品
dedicated plant 专用设备
petrochemical *n.* 石油化学产品
substance *n.* 物质
multipurpose plant 多用途设备
specification *n.* 规格
ingredient *n.* 原料
adhesive *n.* 黏合剂
agrochemical *n.* 农用化学品
dye *n.* 染料
pigment *n.* 颜料
pharmaceutical *n.* 药物
amino acid 氨基酸
high-added-value product 高附加值产品
selectivity *n.* 选择性
side product 副产物
hazardous *adj.* 有危险的
flammable *adj.* 易燃的
cyanide *n.* 氰化物
phosgene *n.* 光气
halogen *n.* 卤素
isocyanate *n.* 异氰酸盐
phosphorous *adj.* 含磷的

Notes

➢ Fine chemicals are complex, single, pure chemical substances. They are produced in limited quantities (up to 1000Mt per year) in multipurpose plants by multistep batch chemical or biotechnological processes. They are based on exacting specifications, are used as starting materials for specialty chemicals, particularly pharmaceuticals, biopharmaceuticals and agrochemicals, and are sold for more than $10/kg.

参考译文：精细化学品是复杂的、单一的、纯的化学物质。它们通过多步间歇反应或生物技术过程在多用途设备中生产且生产的数量有限（每年最多 10 亿吨）。它们基于严格的标准，被用作特种化学品，特别是药品、生物制药和农用化学品的初始原料，且售价超过 10 美元/公斤。

➢ The distinction between fine and specialty chemicals is net. The former is sold on the basis of "what they are"; the latter, on "what they can do."

参考译文：精细化学品和专用化学品之间的区别是净利润。前者是根据“它们是什么”来出售的，后者则是基于“它们能做什么”。

➢ Fine chemicals are often manufactured in multistep conventional syntheses, which results in a high consumption of raw materials and, consequently, large amounts of by-products and wastes.

参考译文：精细化学品通常是通过多步传统合成方法生产的。这种合成方法导致了原料的高消耗，以及随之而来的大量副产物和废物。

➢ On the one hand, chemical and process engineers are dedicated mainly to the engineering part in process development, often neglecting process chemistry, as this might be considered less important for a full-scale plant. On the other hand, synthetic chemists often finish their work with laboratory recipes neglecting needs of the procedures for process development.

参考译文：一方面，化学和工艺工程人员主要关心（化工）工艺开发的工程部分，而常常忽略工艺化学，因为这对整个工厂来讲可能不那么重要；另一方面，合成化学家常常在得到实验室配方后就结束工作，忽略了化工工艺开发程序的需要。

➢ The application of techniques cultivated in semiconductor industries have allowed the creation of a new instrumental platform able to efficiently manipulate, process and analyze molecular reactions on the micrometer to nanometer scale.

参考译文：半导体工业培育的技术可以创造出一种新的仪器平台，该平台能有效地操纵、处理和分析在微米到纳米尺度上的分子反应。

Exercises

1. Discuss the following questions

(1) What are the characteristics of fine chemicals?

(2) What is the definition of microfluidics?

(3) List the classifications of the fine chemicals.

(4) What are the advantages of microfluidic systems?

2. Put the following into Chinese

fine chemical　agrochemical　petrochemical　pigment　ingredient

dedicated plant　selectivity　hazardous　isocyanate　phosgene

3. Put the following into English

通用化学品　药物　染料　黏合剂　氨基酸　副产物

易燃的　氰化物　含磷的　卤素

4. Translate the following into Chinese

(1) The accuracy of analytical methods has increased enormously in the past decades and this has enabled detection of even almost negligible traces of impurities. The consequence is that both regulations and specifications for intermediates and final fine chemicals have become stricter. Therefore, very pure compounds must often be produced with impurities at 10^{-6} or 10^{-9} level. The production of complex molecules in many cases results in mixture containing isomers, including optical isomers. The demand for enantiomeric materials is growing at the expense of their racemic counterparts, driven primarily by the pharmaceutical industry.

(2) For instance, because of its precise controllability and operating stability, microfluidic technology could be employed to prepare monodispersed droplets or micro-bubbles with polydispersity index less than 5%. In addition, the heat and mass transfer coefficients within microchannels are proved to be one to two orders of magnitude higher than those of macroscale reactors and mixers, making uniform mixing, fast extraction, and reaction intensification possible in micromixers, microreactors, and microdispersers, respectively.

UNIT 2 Dyes and Pigments

Colour may be introduced into manufactured articles, for example textiles and plastics, or into a range of colour application media, for example paints and printing inks, for a variety of reasons but the most commonly ultimate purpose is to enhance the appearance and attractiveness of a product and improve its market appeal. The desired colour is generally achieved by the incorporation into the product of coloured compounds referred to as dyes and pigments. The term colourant is frequently used to encompass both types of colouring materials.

1. History of Synthetic Dyes

Until the 1850s, virtually all dyes were obtained from natural sources, most commonly from vegetables, such as plants, trees, and lichens, with a few from insects. Countless attempts have been made to extract dyes from brightly coloured plants and flowers, yet only a dozen or so natural dyes found widespread use. Undoubtedly most attempts failed because most natural dyes are not highly stable and occur as components of complex mixtures, the successful separation of which would be unlikely by the crude methods employed in ancient times. Nevertheless, studies of these dyes in the 1800s provided a base for development of synthetic dyes, which dominated the market by 1900.

Fig.4.3 Photographs of William Henry Perkin and Malva

The first synthetic dye, mauveine, was discovered by Perkin in 1856 (Fig.4.3). Hence the dyestuffs industry can rightly be described as mature. However, it remains a vibrant, challenging industry requiring a continuous stream of new products because of the quickly changing world in which we live. The early dye industry saw the discovery of the principal dye chromophores (the basic arrangement of atoms responsible for the colour of a dye). Indeed, apart from one or two notable exceptions, all the dye types used today were discovered in the 1800s. The introduction of

音频

the synthetic fibres, nylon, polyester, and polyacrylonitrile during the period 1930~1950, produced the next significant challenge. The discovery of reactive dyes in 1954 and their commercial launch in 1956 heralded a major breakthrough in the dyeing of cotton; intensive research into reactive dyes followed over the next two decades and, indeed, is still continuing today. The oil crisis in the early 1970s, which resulted in a steep increase in the prices of raw materials for dyes, created a drive for more cost-effective dyes, both by improving the efficiency of the manufacturing processes and by replacing tinctorially weak chromophores, such as anthraquinone, with tinctorially stronger chromophores, such as (heterocyclic) azo and benzodifuranone.

2. Distinction between Dyes and Pigments

For an appreciation of the chemistry of colour application, it is of fundamental importance that the distinction is made between dyes and pigments as quite different types of colouring materials. Dyes and pigments are both commonly supplied by the manufacturers as coloured powders. Indeed, as the discussion of their molecular structures contained in this unit will illustrate, the two groups of colouring materials may often be chemically quite similar. However, they are distinctly different in their properties and especially in the way they are used. Dyes and pigments are distinguished on the basis of their solubility characteristics: Essentially, dyes are soluble, pigments are insoluble.

Dyes are almost invariably applied to the textile materials from an aqueous medium, so that they are generally required to dissolve in water. Frequently, as is the case for example with acid dyes, direct dyes, cationic dyes and reactive dyes, (some of the application classes of textile dyes discussed later in this unit) they dissolve completely and very readily in water. This is not true, however, of every application class of textile dyes. Disperse dyes for polyester fibres, for example, are only sparingly soluble in water and are applied as a fine aqueous dispersion. Vat dyes, an important application class of dyes for cellulosic fibres, are completely insoluble materials but they are converted by a chemical reduction process to a water-soluble form which may then be applied to the fibre. There are also a wide range of non-textile applications of dyes, many of which have emerged in recent years as a result of developments in the electronic and reprographic industries. For many of these applications, solubility in specific organic solvents rather than in water is of importance.

Pigments are colouring materials which are required to be completely insoluble in the medium into which they are incorporated. The traditional applications of pigments are in paints, printing inks and plastics, although they are also used more widely, for example, in the colouration of building materials, such as concretes and cements, and in ceramics and glasses. Pigments are applied into a medium by a dispersion process, which reduces the clusters of solid particles into a more finely-divided form, but they do not dissolve in the medium. They remain as solid particles held in place mechanically, usually in a matrix of a solid polymer. A further distinction between dyes and pigments is that while dye molecules are designed to be attracted strongly to the polymer molecules which make up the textile fibre, pigment molecules are not required to exhibit such affinity for their medium. Pigment molecules are, however, required to be attracted strongly to one another in their solid crystal lattice structure in order to resist dissolving in solvents.

3. Classifications of Dyes

Dyes may be classified according to chemical structure or by their usage or application method. Neither of these two categories can be used with the exclusion of the other one, and overlap is often inevitable. This unit is devoted to the chemical chromophores of dyes. The two overriding trends in traditional colourants research for many years have been improved cost-effectiveness and increased technical excellence. Improved cost-effectiveness usually means replacing tinctorially weak dyes such as anthraquinones, until recently the second largest class after the azo dyes, with tinctorially stronger dyes such as heterocyclic azo dyes, triphendioxazines, and benzodifuranones.

Direct or substantive dyes are applied to the fabric from a hot aqueous solution of the dye. Under these conditions, the dye is more soluble and the wettability of natural fibres is increased, improving the transport of dye molecules into the fabric. In many cases, the fabric is pretreated with metallic salts or mordants to improve the fastness and to vary the colour produced by a given dye.

4. Azo Dyes

The azo dyes are by far the most important class, accounting for over 50% of all commercial dyes, and having been studied more than any other class. The molecular structures of azo dyes contain one or several azo groups —N═N— which connect aromatic rings (Fig.4.4). Depending on the number of such groups, the dyes are called mono-, dis-, tris-, or polyazo dyes. Usually, azo dyes contain substituted or unsubstituted —NH_2, —OH, —NO_2, —Cl, —SO_3H, —COOH, and other groups on the aromatic rings. The presence of acid groups ensures the water solubility of the dyes.

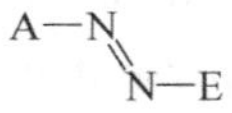

A—N
   ╲╲
     N—E

Fig.4.4 General chemical structure of azo dyes

In monoazo dyes, the most important type, the A group often contains electron-accepting substituents, and the E group contains electron-donating substituents, particularly hydroxyl and amino groups. If the dyes contain only aromatic groups, such as benzene and naphthalene, they are known as carbocyclic azo dyes. If they contain one or more heterocyclic groups, the dyes are known as heterocyclic azo dyes. The simplest monoazo dyes are usually yellow, orange, or red in colour. An increase in the number of azo groups, a substitution of phenyl radicals with naphthyl radicals, and an increase in the number of oxy- and amino- groups will intensify the colour. Depending on their structure and the character of their interaction with textiles, azo dyes are divided into several groups: basic, acid, direct, mordant, ice-colour, active, and others. Azo dyes are used mainly for colouring textiles but also for colouring leather, paper, rubber, and certain plastics.

5. Anthraquinone Dyes

Once the second most important class of dyes, it also includes some of the oldest dyes. They have been found in the wrappings of mummies dating back over 4000 years. In contrast to the azo dyes, which have no natural counterparts, all the important natural red dyes are anthraquinones. However, the importance of anthraquinone dyes has declined due to their low cost-effectiveness.

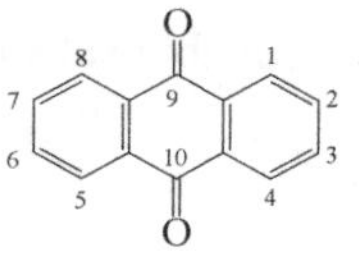

Fig.4.5 Structure of 9,10-anthraquinone

Anthraquinone dyes are based on 9,10-anthraquinone, which is essentially colourless (Fig.4.5). To produce commercially useful dyes, strongly electron-donating groups such as amino or hydroxyl are introduced into one or more of the four positions (1, 4, 5, and 8). The most common substitution patterns are 1,4-, 1,2,4-, and 1,4,5,8-. To optimize the properties, primary and secondary amino groups (not tertiary) and hydroxyl groups are employed. These ensure the maximum degree of JT-orbital overlap, enhanced by intramolecular hydrogen-bonding, with minimum steric hindrance.

6. Indigoid Dyes

Indigoid dyes represent one of the oldest known classes of organic dyes. For 5000 years, they have been used for dyeing textiles such as wool, linen, and cotton. For example, 6,6'-dibromoindigo is Tyrian Purple, the dye made famous by the Romans. Tyrian Purple was so expensive that only the very wealthy were able to afford garments dyed with it. Indeed, the phrase “born to the purple” is still used today to denote wealth. Although many indigoid dyes have been synthesized, only indigo itself is of any major importance today. Indigo is the blue used almost exclusively for dyeing denim jeans and jackets and is held in high esteem because it fades to give progressively paler blue shades.

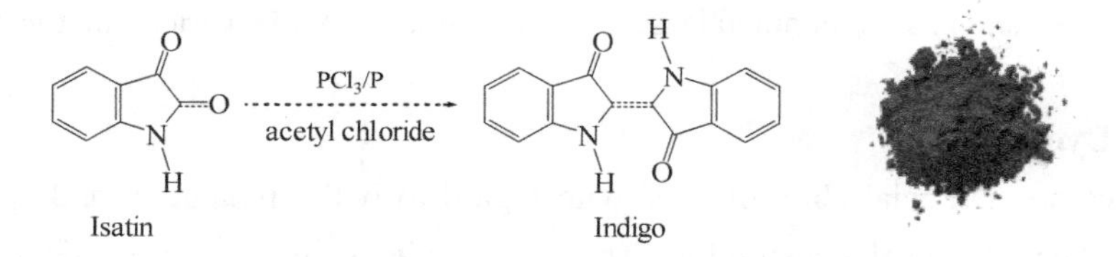

Fig.4.6 Synthetic routine of indigo from isatin

Adolf von Baeyer reported the chemical structure of indigo in 1883, having already prepared the first synthetic indigo from isatin in 1870. Until then, isatin had only been obtainable by oxidation of indigo. It was not until 1878 that von Baeyer first synthesized it from phenylacetic acid, making a complete synthesis of indigo possible (Fig.4.6).

7. Cationic Dyes

Cationic dyes carry a positive charge in their molecule. The salt-forming counterion in most cases is the colourless anion of a low molecular mass inorganic or organic acid. Many of these dyes can be converted into water-insoluble dye bases by addition of alkali. For this reason, they were formerly called basic dyes, although still in use today, the term should be abandoned. Cationic dyes were used initially for dyeing of silk, leather, paper, and cotton mordanted with tannin, as well as for the production of ink and copying paper in the office supplies industry. Their high brilliance and colour strength partly compensated for their poor lightfastness on these materials. With the development of synthetic fibres, their most suitable substrates were found, and their importance for dyeing textiles increased greatly.

8. Phthalocyanine Dyes

The term phthalocyanine was first used by R. P. Linstead in 1933 to describe a class of organic dyes, whose colours range from reddish blue to yellowish green. The name phthalocyanine originates from the Greek terms naphtha for mineral oil and cyanine for dark blue. In the 1930s, Linstead et al. elucidated the structure of phthalocyanine and its metal complexes. The basic structure is represented by phthalocyanine itself (Fig.4.7).

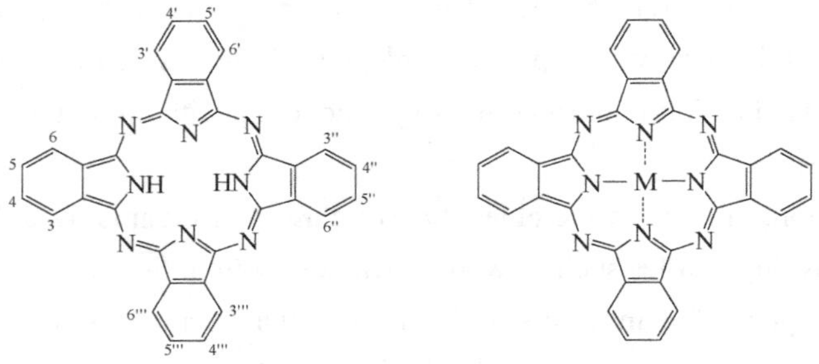

Fig.4.7 Chemical structure of phthalocyanine and metal phthalocyanine

Phthalocyanine forms complexes with numerous metals of the periodic table. A large number of complexes with various elements are known. Metal phthalocyanines and compounds with metalloids such as B, Si, Ge, and As or nonmetals such as P display a wide variety in their coordination chemistry.

9. Sulfur Dyes

Sulfur dyes are a special class of dyes with regard to both preparation and application, and knowledge of their chemical constitution. They are made by heating aromatic or heterocyclic compounds with sulfur or species that release sulfur. Sulfur dyes are classified by method of preparation as sulfur bake, polysulfide bake, and polysulfide melt dyes. Sulfur dyes are not well-defined chemical compounds but mixtures of structurally similar compounds, most of which contain various amounts of both heterocyclic and thiophenolic sulfur.

On oxidation, the monomeric dye molecules crosslink into large molecules by forming disulfide bridges. When a sulfur dye is dissolved by heating with aqueous sodium sulfide solution, the disulfide groups are cleaved to mercapto groups. In this leuco form, the sulfur dye is applied substantively to cotton and other cellulosic fibres. The dyeing is oxidized after rinsing, and the dye molecules recrosslink on the fibre through disulfide groups. Pseudo sulfur dyes are dyes that can be applied in the same way as sulfur dyes but are not prepared by classical sulfurization.

10. Metal-complex Dyes

Metal-complex dyes are coordination compounds in which a metal ion is linked to one or more ligands containing one or more electron-pair donors. Ligands with one and more donor groups are called mono-, di-, trifunctional ligands, etc. Coordination of two or more of the donor groups of such ligands to the same metal atom leads to di-, tri-, or tetradentate chelation, etc. Other names for these ligands are thus chelating agents or chelators. The metal complexes of these ligands are called chelates. The metals in metal-complex dyes are predominantly chromium and copper, and to a lesser extent cobalt, iron, and nickel.

The first use of metal-complex dyes was the process of mordant dyeing, a method that can be traced to the middle ages. Here the textile fabrics to be dyed were impregnated with a solution of salts of a metal such as aluminum, iron, chromium, or tin and then treated with a naturally occurring colourant containing a chelating system to achieve metallization within the fibre. The mordant dyeing leads to a bathochromic shift in colour, albeit with duller hue, and to improve resistance to light and washing. In the 1940s, the so-called after chrome method was developed and gained commercial importance, particularly in the dyeing of wool in dark shades with high fastness. In this process, a chromium complex was formed on the fibre by first dyeing with a metallizable dye followed by after treating the dyed fabrics with sodium or potassium dichromate ($M_2Cr_2O_7 \cdot 2H_2O$) or chromate (M_2CrO_4).

11. Reactive Dyes

Reactive dyes have been very popular for the dyeing and printing of cellulosic fibre for many years. The fibre-reactive dyes were first introduced as the Procion dyes of ICI in 1956 for the production of fast brilliant colours on cellulosic materials by continuous dyeing methods. Reactive dyes were immediately proved attractive to dyers due to the bright colours and the excellent fastness properties of this new dye class. This immediate interest was further increased when batch dyeing methods for these dyes were developed. They are water-soluble anionic dyes and various physical forms of these dyes are available such as pourable granules, finished powders and highly concentrated aqueous solutions, whereas in the past dyes were mainly marketed as powders. There are also commercially available fibre-reactive dyes for protein and polyamide fibres.

Reactive dyes consist of four parts:

① The chromophore or the chromophoric part, which contributes colour to the dye.

② The reactive system, which enables the dye to react with the substrate. This part can also react with water molecules present in the dyebath, a phenomenon called hydrolysis of the reactive dyes which is not a desirable reaction during dyeing.

③ A bridging unit that joins the reactive system to the chromophoric part.

④ The solubilising group(s) attached to the chromophoric grouping confers water solubility to the dye.

In a few cases, the reactive part remains attached directly to the dye. Substituent groups and the nature of the bridging unit affect the reactivity and dyeing characteristics of dyes. The majority of reactive dyes are chemically of the azo class, although anthraquinone-based reactive dyes are also available. Depending on the type of reaction, the reactive dyes are broadly divided into two categories: dyes reacting through the nucleophilic substitution reaction and dyes reacting through the nucleophilic addition reaction.

Triazine-based reactive groups react through the nucleophilic substitution reaction. They are mostly chlorotriazines of either the dichloro or monochloro type. The reactivity of dichlorotriazine dyes is higher than monochloro dyes, so they require a lower temperature and a milder alkali for their application. These dyes are known as cold brand reactive dyes. The trade names of some cold brand dyes are Reactofix M (Jay Synth), Procion M (Atul) and Procion MX (Zeneca). The general chemical structure of these dyes is shown in Fig.4.8.

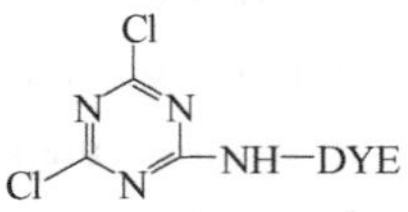

Fig.4.8 General structure of cold brand reactive dye

Monochlorotriazine dyes require a higher temperature and a stronger alkali application than the dichlorotriazine dyes. Monochlorotriazine dyes are available on the market in different trade names such as Procion H (Atul), Reactofix H (Jay Synth), Amaryl X (Amar Dye Chem), etc. The letter "M" and "H" in the trade names of dichlorotriazine and monochlorotriazine dyes respectively stand for cold brand dyes and hot brand dyes. The general structure of monochlorotriazine hot brand reactive dyes is presented in Fig.4.9.

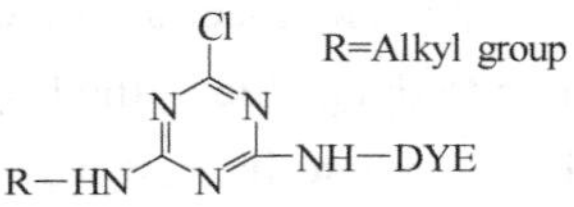

Fig.4.9 General structure of hot brand reactive dye

The reaction of triazine dyes with cellulose is shown in Fig.4.10. The chlorine of the triazine ring is replaced by O—cellulose. Thus the dyes form a covalent bond with the fibre, which accounts for the high wash-fastness of reactive dyes. The reactive dyes can react with water molecules as well. The reaction of the dyes with water is called hydrolysis. The hydrolysed dyes lose their power to react with the fibre. These dyes remain loosely held on the fibre and if they are not properly removed through thorough soaping, the wash-fastness of the dyes are deteriorated. The nature of the electrolyte used plays an important role in the absorption of the hydrolysed dye. Shrivastava (1979) studied the effect of various electrolytes on the absorption of the hydrolysed dye on cotton and found that their relative efficiencies were in the order: ammonium chloride > ammonium sulphate > sodium chloride > lithium chloride ~ magnesium chloride. Ciba Geigy developed Cibacron F dyes where fluorine is used in place of chlorine as the leaving group.

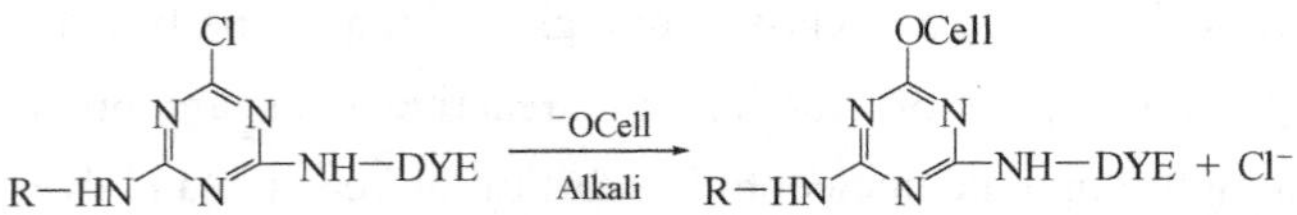

Fig.4.10 Reaction of cotton cellulose with reactive dye

Dyes having an acrylamido-based reactive system are developed by BASF (Primazine dyes) and Ciba Geigy (Lanasyn/Lanasol dyes). The general structure of these dyes is D—NHCO—CHX=CH_2. The carbonyl group is a less powerful electron-withdrawing group so the reactivity of these dyes is less compared with vinyl sulphone dyes. Dyes containing hetero-bifunctional groups like Sumifix dyes (Sumitomo Chemical Co.), which contain both chlorotriazine and vinyl sulphone reactive groups are also very popular. The bond between vinyl sulphone and cellulose is very strong and is stable to acid hydrolysis. Also the substantivity of hydrolysed dyes is very poor and so unfixed hydrolysed dyes can be easily washed off. The advantage of the chlorotriazine group is that it increases the substantivity of dyes and thus improves the degree of exhaustion and fixation of dyes.

Words and Expressions

colourant *n.* 着色剂
chromophore *n.* 发色团
affinity *n.* 亲和力
fibre *n.* 纤维
polyester *n.* 聚酯纤维
polyacrylonitrile *n.* 聚丙烯腈
azo dye 偶氮染料
intramolecular hydrogen-bonding 分子内氢键
indigoid dye 靛蓝类染料
indigo *n.* 靛蓝
isatin *n.* 靛红
phthalocyanine *n.* 酞菁
thiophenolic sulfur 噻吩硫
chelate *n.* 螯合物
heterocyclic group 杂环基团
anthraquinone *n.* 蒽醌
cationic *adj.* 阳离子的
substantivity *n.* 直接性；直染性
cellulosic *adj.* 有纤维质的
reactive dye 活性染料
hydrolysis *n.* 水解作用
substituent group 取代基
solubilizing *adj.* 增溶的
nucleophilic *adj.* 亲核的

Notes

➢ The traditional applications of pigments are in paints, printing inks and plastics, although they are also used more widely, for example, in the colouration of building materials, such as concretes and cements, and in ceramics and glasses.

参考译文：颜料的传统应用是在油漆、印刷油墨和塑料中，尽管它们也被广泛使用在建筑材料（如混凝土和水泥）的着色中，以及在陶瓷和玻璃中。

➢ Vat dyes, an important application class of dyes for cellulosic fibres, are completely insoluble materials but they are converted by a chemical reduction process to a water-soluble form which may then be applied to the fibre.

参考译文：还原染料是纤维素纤维染料的一个重要应用类别，是完全不溶性物质，但它们可以通过化学还原过程转化为可溶于水的形式，然后被应用于纤维。

➢ Improved cost-effectiveness usually means replacing tinctorially weak dyes such as anthraquinones, until recently the second largest class after the azo dyes, with tinctorially stronger dyes such as heterocyclic azo dyes, triphendioxazines, and benzodifuranones.

参考译文：提高成本效益通常意味着用着色较强的染料（如杂环偶氮染料、三苯二噁嗪和苯二氮呋喃酮）取代着色较弱的染料，如蒽醌类（直到最近，蒽醌类成为仅次于偶氮染料的第二大类染料）。

➢ On oxidation, the monomeric dye molecules crosslink into large molecules by forming disulfide bridges. When a sulfur dye is dissolved by heating with aqueous sodium sulfide solution, the disulfide groups are cleaved to mercapto groups.

参考译文：在氧化过程中，单体染料分子通过二硫桥联形成大分子。当硫染料与硫化钠水溶液加热溶解时，二硫基则裂解为巯基。

➢ Cationic dyes were used initially for dyeing of silk, leather, paper, and cotton mordanted with tannin, as well as for the production of ink and copying paper in the office supplies industry. Their high brilliance and colour strength partly compensated for their poor lightfastness on these materials.

参考译文：阳离子染料最初用于丝绸、皮革、纸张和棉的染色，并用鞣质进行媒染，以及办公用品中墨水和复印纸的生产。它们的高亮度和色强部分弥补了它们在这些材料上的不耐光性。

Exercises

1. Discuss the following questions

(1) What is the difference between dyes and pigments?

(2) How to intensify the colour of azo dyes?

(3) Which groups can azo dyes be divided into?

(4) How many parts do reactive dyes consist of and what are they?

(5) How to divide reactive dyes?

2. Put the following into Chinese

indigo   chelate   heterocyclic group   ionization   cellulosic

azo   polyester   intramolecular hydrogen-bonding   fibre   hydrolysis

3. Put the following into English

染料   颜料   发色团   活性染料   增溶的

蒽醌   酞菁   亲核的   亲和力   阳离子的

4. Translate the following into Chinese

(1) For an appreciation of the chemistry of colour application, it is of fundamental importance that the distinction is made between dyes and pigments as quite different types of colouring materials. Dyes and pigments are both commonly supplied by the manufacturers as coloured powders. Indeed, as the discussion of their molecular structures contained in this unit will illustrate,

the two groups of colouring materials may often be chemically quite similar. However, they are distinctly different in their properties and especially in the way they are used. Dyes and pigments are distinguished on the basis of their solubility characteristics: Essentially, dyes are soluble, pigments are insoluble.

(2) The simplest monoazo dyes are usually yellow, orange, or red in colour. An increase in the number of azo groups, a substitution of phenyl radicals with naphthyl radicals, and an increase in the number of oxy- and amino- groups will intensify the colour. Depending on their structure and the character of their interaction with textiles, azo dyes are divided into several groups: basic, acid, direct, mordant, ice-colour, active, and others. Azo dyes are used mainly for colouring textiles but also for colouring leather, paper, rubber, and certain plastics.

UNIT 3 Surfactants

1. Surfactants Adsorb at Interfaces

Surfactant is an abbreviation for surface active agent, which literally means active at a surface. In other words, a surfactant is characterized by its tendency to adsorb at surfaces and interfaces. The term interface denotes a boundary between any two immiscible phases, the term surface indicates that one of the phases is a gas, usually air.

The driving force for a surfactant to adsorb at an interface is to lower the free energy of that phase boundary. The interfacial free energy per unit area represents the amount of work required to expand the interface. The term interfacial tension is often used instead of interfacial free energy per unit area. Thus, the surface tension of water is equivalent to the interfacial free energy per unit area of the boundary between water and the air above it. When that boundary is covered by surfactant molecules, the surface tension (or the amount of work required to expand the interface) is reduced. The denser the surfactant packing at the interface, then the larger the reduction in surface tension.

Surfactants may adsorb at all of the five types of interfaces. Here, the discussion will be restricted to interfaces involving a liquid phase. The liquid is usually, but not always water. Examples of different interfaces and products in which these interfaces are important are given in Table 4.1.

Table 4.1 Examples of interfaces involving a liquid phase

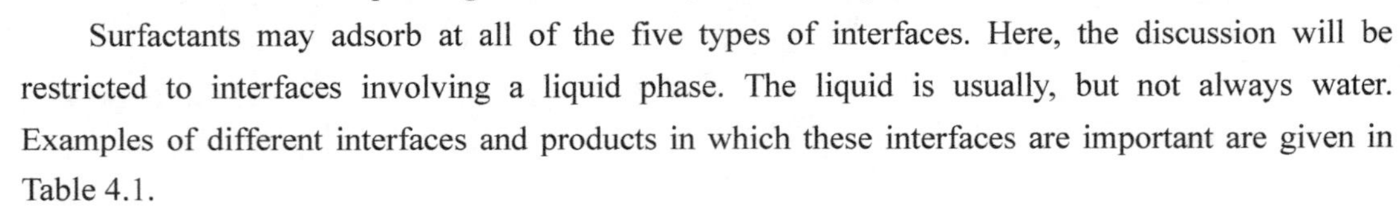

Interface	Type of System	Product
Solid-liquid	Suspension	Solvent-borne paint
Liquid-liquid	Emulsion	Milk, cream
Liquid-vapour	Foam	Shaving cream

In many formulated products, several types of interfaces are present at the same time. Water-based paints and paper coating colours are examples of familiar but, from a colloidal point of view, very complicated systems containing both solid-liquid (dispersed pigment particles) and liquid-liquid (latex or other binder droplets) interfaces. In addition, foam formation is a common (but unwanted) phenomenon at the application stage. All of the interfaces are stabilized by surfactants. The total interfacial area of such a system is immense: The oil-water and solid-water interfaces of one liter of paint may cover several football fields.

As mentioned above, the tendency to accumulate at interfaces is a fundamental property of a surfactant. In principle, the stronger the tendency, then the better the surfactant. The degree of surfactant concentration at a boundary depends on the surfactant structure and also on the nature of the two phases that meet at the interface. Therefore, there is no universally good surfactant, suitable for all uses. The choice will depend on the application. A good surfactant should have low solubility in the bulk phases. Some surfactants (and several surface active macromolecules) are only soluble at the oil-water interface. Such compounds are difficult to handle but are very efficient in reducing the interfacial tension.

There is, of course, a limit to the surface and interfacial tension lowering effect by the surfactant. In the normal case, that limit is reached when micelles start to form in bulk solution. Table 4.2 illustrates what effective surfactants can do in terms of lowering surface and interfacial tension. The values given are typical of what are attained by normal light-duty liquid detergents.

Table 4.2 Typical values of surface and interfacial tensions

Surfactant	Tension/($mN \cdot m^{-1}$)
Air–water	72~73
Air–10% aqueous NaOH	78
Air–aqueous surfactant solution	40~50
Aliphatic hydrocarbon–water	28~30
Aromatic hydrocarbon–water	20~30
Hydrocarbon–aqueous surfactant solution	1~10

2. Surfactants Aggregate in Solution

As discussed above, one characteristic feature of surfactants is their tendency to adsorb at interfaces. Another fundamental property of surface-active agents is that unimers in solution tend to form aggregates, so-called micelles. (The free or unassociated surfactant is referred to in the literature either as "monomer" or "unimer". In this text, we will use "unimer" and the term "monomer" will be restricted to the polymer building block.) Micelle formation, or micellization, can be viewed as an alternative mechanism to adsorption at the interfaces for removing hydrophobic groups from contacting with water, thereby reducing the free energy of the system. It is an important phenomenon since surfactant molecules behave very differently when present in micelles than as free unimers in solution. Only surfactant unimers contribute to surface and interfacial tension lowering and dynamic phenomena, such as wetting and foaming, are governed by the concentration of free unimers in solution. The micelles may be seen as a reservoir for surfactant unimers. The exchange rate of a surfactant molecule between micelle and bulk solution may vary by many orders of magnitude depending on the size and structure of the surfactant.

Micelles are already generated at very low surfactant concentrations in water. The concentration at which micelles start to form is called the critical micelle concentration, or CMC, and is an important characteristic of a surfactant. A CMC of $1 mmol \cdot L^{-1}$, a reasonable value for an ionic surfactant, means that the unimer concentration will never exceed this value, regardless of the amount of surfactant added to the solution.

3. Surfactants are Amphiphilic

The name amphiphile is sometimes used synonymously with surfactant. The word is derived from the Greek word amphi, meaning both, and the term relates to the fact that all surfactant molecules consist of at least two parts, one which is soluble in a specific fluid (the lyophilic part) and one which is insoluble (the lyophobic part). When the fluid is water, one usually talks about the hydrophilic and hydrophobic parts, respectively. The hydrophilic part is referred to as the head group and the hydrophobic part as the tail (see Fig.4.11).

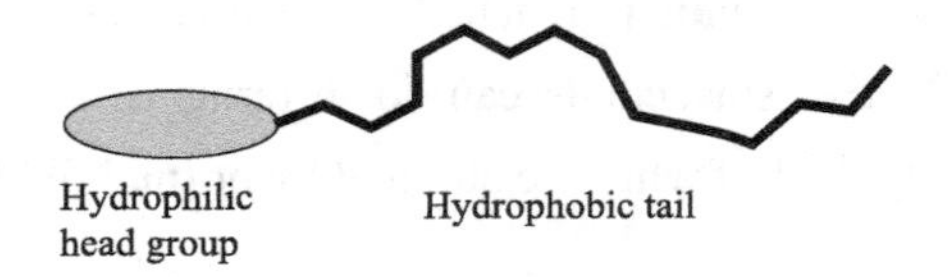

Fig.4.11 Schematic illustration of a surfactant

In a micelle, the surfactant hydrophobic group is directed towards the interior of the cluster and the polar head group is directed towards the solvent. The micelle, therefore, is a polar aggregate of high water solubility and without much surface activity. When a surfactant adsorbs from an aqueous solution at a hydrophobic surface, it normally orients its hydrophobic group towards the surface and exposes its polar group to the water. The surface has become hydrophilic and, as a result, the interfacial tension between the surface and water has been reduced.

The hydrophobic part of a surfactant may be branched or linear alkyl chains. The polar head group is usually, but not always, attached at one end of the alkyl chain. The length of the chain is in the range of 8~18 carbon atoms. The degree of chain branching, the position of the polar group and the length of the chain are parameters of importance for the physicochemical properties of the surfactant.

The polar part of the surfactant may be ionic or non-ionic and the choice of polar group determines the properties to a large extent. For non-ionic surfactants, the size of the head group can be varied at will. For the ionics, the size is more or less a fixed parameter. The relative size of the hydrophobic and polar groups, not the absolute size of either of the two, is decisive in determining the physicochemical behaviour of a surfactant in water.

4. Surfactants are Classified by the Polar Head Group

The primary classification of surfactants is made on the basis of the charge of the polar head group. It is common practice to divide surfactants into the classes of anionics, cationics, non-ionics and zwitterionics. Surfactants belonging to the latter class contain both an anionic and a cationic charge under normal conditions. In the literature, they are often referred to as amphoteric surfactants but the term "amphoteric" is not always correct and should not be used as synonymous to zwitterionic. An amphoteric surfactant is one that, depending on pH, can be either cationic, zwitterionic or anionic. Among normal organic substances, simple amino acids are well-known examples of amphoteric compounds. Many so-called zwitterionic surfactants are in this category. However, other zwitterionic surfactants retain one of the charges over the whole pH range. Compounds with a quaternary ammonium as the cationic group are examples of this. Consequently, a surfactant that contains a carboxylate group and a quaternary ammonium group, a not uncommon combination, is zwitterionic unless the pH is very low, but is not an amphoteric surfactant.

Most ionic surfactants are monovalent but there are also important examples of divalent anionic amphiphiles. For the ionic surfactants, the choice of counterion plays a role in the physicochemical properties. Most anionic surfactant shave sodium as a counterion but other cations, such as lithium, potassium, calcium and protonated amines, are used as surfactant counterions for special purposes. The counterion of cationic surfactants is usually a halide or methyl sulfate.

The hydrophobic group is normally a hydrocarbon but may also be a polydimethylsiloxane or a fluorocarbon. The two latter types of surfactants are particularly effective in non-aqueous systems.

5. Anionics

Carboxylate, sulfate, sulfonate and phosphate are the polar groups found in anionic surfactants. Fig.4.12 shows structures of the more common surfactant types belonging to this class.

Anionics are used in greater volume than any other surfactant class. A rough estimate of the worldwide surfactant production is 10 million tons per year, out of which approximately 60% are anionics. One main reason for their popularity is the ease and low cost of manufacture. Anionics are used in most detergent formulations and the best detergency is obtained by alkyl and alkylarye chains in the C_{12}~C_{18} range.

The counterions most commonly used are sodium, potassium, ammonium, calcium and various protonated alkyl amines. Sodium and potassium impart water solubility, whereas calcium and magnesium promote oil solubility. Amine/ alkanol amine salts give products with both oil and water solubility.

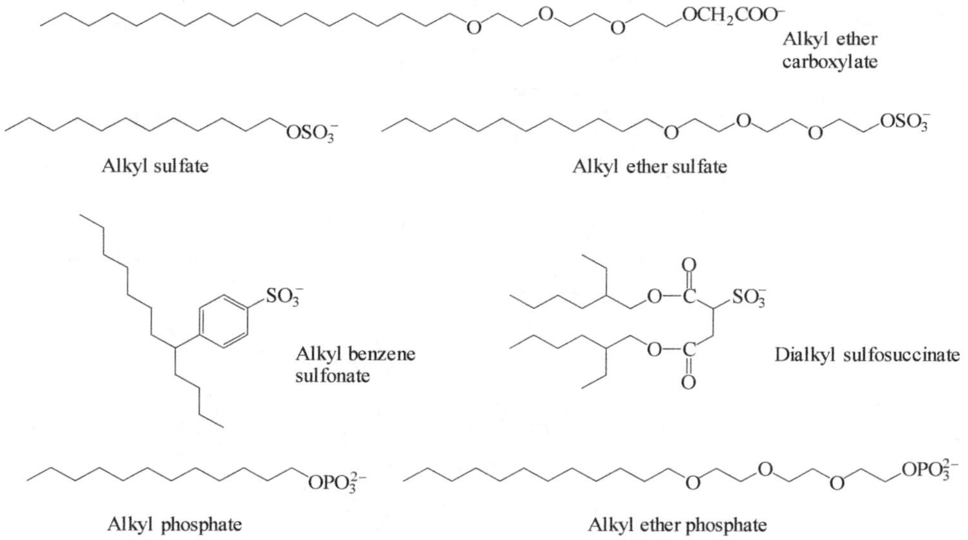

OCH_2COO^-
Alkyl ether carboxylate
OSO_3^-
Alkyl sulfate
OSO_3^-
Alkyl ether sulfate
SO_3^-
Alkyl benzene sulfonate
SO_3^-
Dialkyl sulfosuccinate
OPO_3^{2-}
Alkyl phosphate
OPO_3^{2-}
Alkyl ether phosphate

Fig.4.12 Chemical structures of representative anionics

6. Non-ionics

Non-ionic surfactants have either a polyether or a polyhydroxyl unit as the polar group. In the vast majority of non-ionics, the polar group is a polyether consisting of oxyethylene units, made by the polymerization of ethylene oxide. Strictly speaking, the prefix "poly" is a misnomer. The typical number of oxyethylene units in the polar chain is five to ten, although some surfactants, e.g. dispersants, often have much longer oxyethylene chains. Fig.4.13 gives structures of the more common non-ionic surfactants. As mentioned below, a commercial oxyethylene-based surfactant consists of a very broad spectrum of compounds, broader than most other surfactant types. Fatty acid ethoxylates constitute particularly complex mixtures with high amounts of poly(ethylene glycol) and fatty acid as by-products. The single most important type of non-ionic surfactant is fatty alcohol ethoxylates. They are used in liquid and powder detergents as well as in a variety of industrial applications. They are particularly useful to stabilize oil-in-water emulsions. Fatty alcohol ethoxylates can be regarded as hydrolytically stable in the pH range of 3~11. They undergo a slow oxidation in air, however, and some oxidation products, e.g. aldehydes and hydroperoxides, are more irritating to the skin than the intact surfactant.

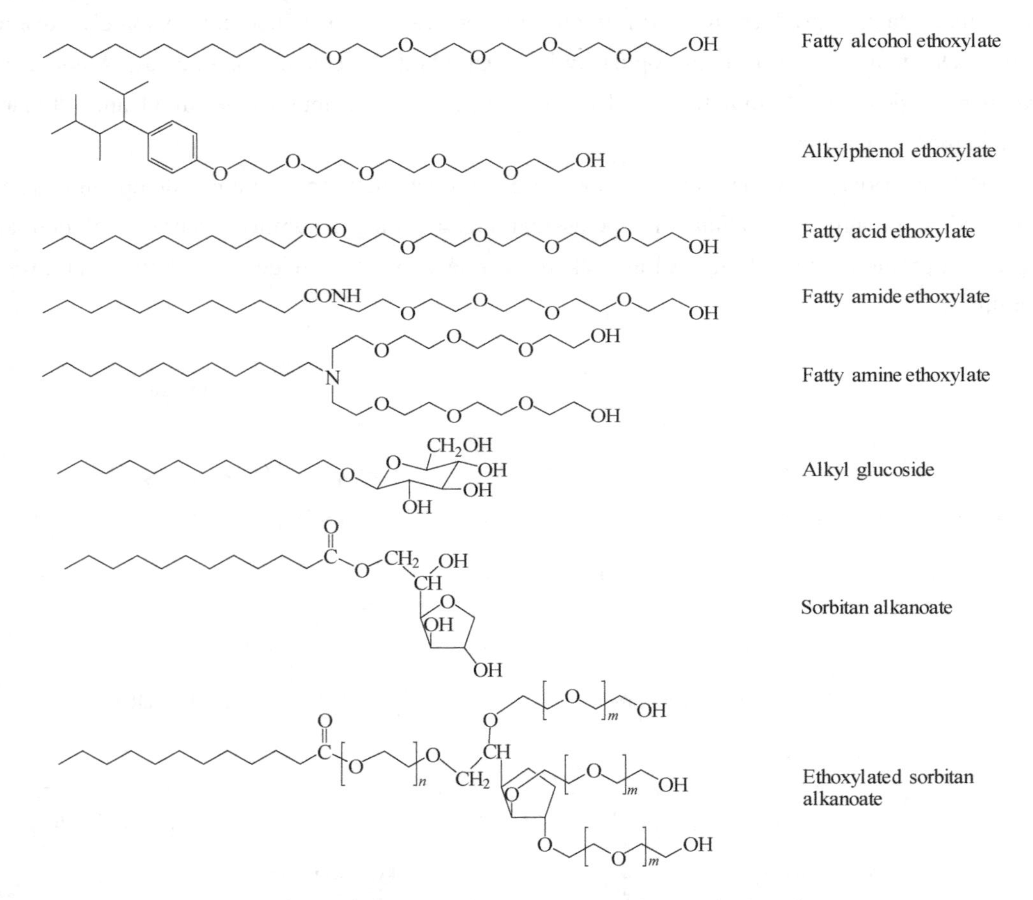

Fig.4.13 Chemical structures of some representative non-ionic surfactants

7. Cationics

The vast majority of cationic surfactants are based on the nitrogen atom carrying the cationic charge. Both amine and quaternary ammonium-based products are common. The amines only function as a surfactant in the protonated state. Therefore, they cannot be used at high pH. Quaternary ammonium compounds, "quats", are not pH sensitive. Non-quaternary cationics are also much more sensitive to polyvalent anions. The ethoxylated amines (see Fig.4.13) possess characteristics of both cationics and non-ionics. The longer the polyoxyethylene chain, then the more non-ionic the character of this surfactant type.

Fig.4.14 shows the structures of some typical cationic surfactants. The ester "quat" represents a new, environmentally friendly type which to a large extent has replaced dialkyl quats as textile softening agents.

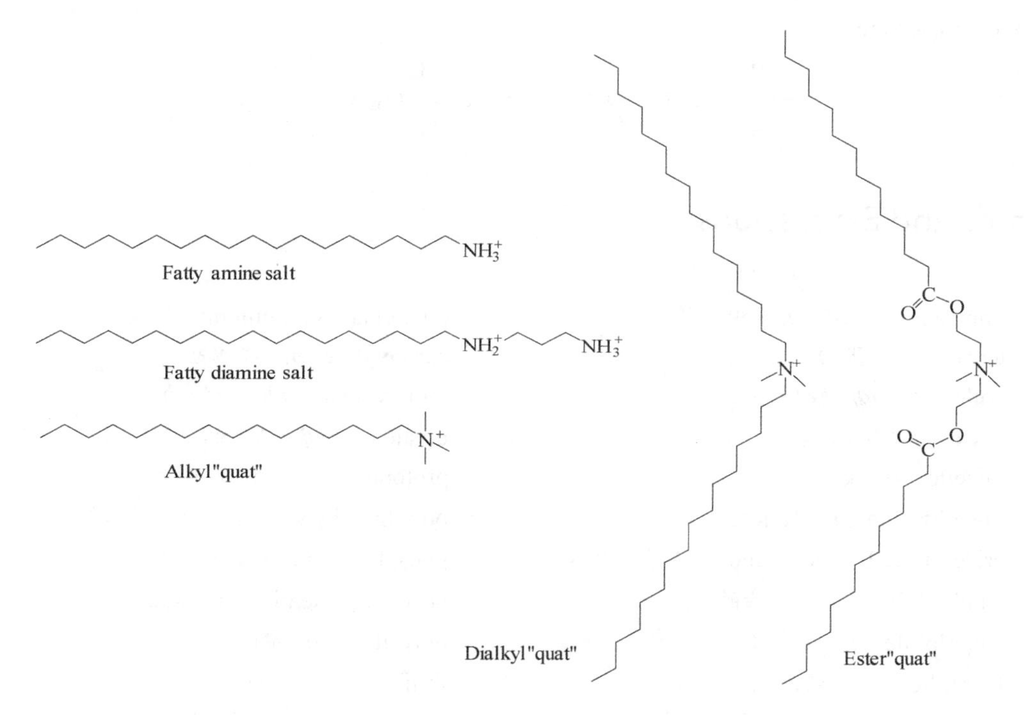

Fig.4.14 Chemical structures of some representative cationics

8. Zwitterionics

Zwitterionic surfactants contain two charged groups of different signs. Whereas the positive charge is almost invariably ammonium, the source of negative charge may vary, although carboxylate is by far the most common. Zwitterionics are often referred to as amphoterics. An amphoteric surfactant is one that changes from net cationic via zwitterionics to net anionic on going from low to high pH. Neither the acid nor the base site is permanently charged, i.e. the compound is only zwitterionic over a certain pH range.

The change in charge with pH of the truly amphoteric surfactants naturally affects properties such as foaming, wetting, detergency, etc. These will all depend strongly on the solution pH. At the isoelectric point, the physicochemical behaviour often resembles that of non-ionic surfactants. Below and above the isoelectric point, there is a gradual shift towards the cationic and anionic character, respectively. Surfactants based on sulfate or sulfonate to give a negative charge remain zwitterionic down to very low pH values due to the very low pK_a values of monoalkyl sulfuric acid and alkyl sulfonic acid, respectively.

Common types of zwitterionic surfactants are *N*-alkyl derivatives of simple amino acids, such as glycine (NH_2CH_2COOH), betaine ($(CH_3)_3\overset{+}{N}CH_2COO^-$) and amino propionic acid ($NH_2CH_2CH_2COOH$). They are usually not prepared from the amino acid, however, by reacting a long-chain amine with sodium chloroacetate or a derivative of acrylic acid, giving structures with one and two carbons, respectively, between the nitrogen and the carboxylate group. As an example, a typical betaine surfactant is prepared by reacting an alkyldimethyl amine with sodium monochloroacetate.

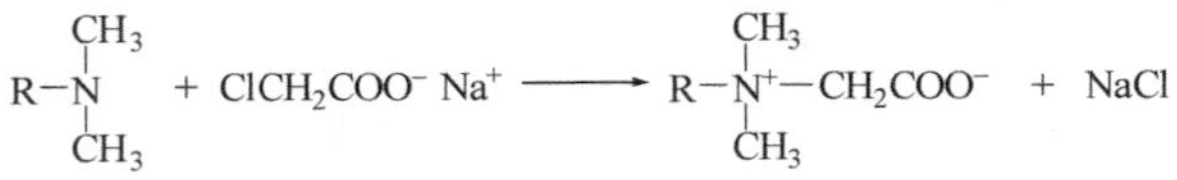

$$R-N(CH_3)_2 + ClCH_2COO^- Na^+ \longrightarrow R-N^+(CH_3)_2-CH_2COO^- + NaCl$$

Words and Expressions

immiscible *adj.* 互不相溶的
tension *n.* 张力
colloidal *adj.* 胶体的
latex *n.* 乳胶；乳液
micelle *n.* 胶束
micellization *n.* 胶束化
critical micelle concentration 临界胶束浓度
amphiphilic *adj.* 两亲的
amphiphile *n.* 两亲物
lyophilic *adj.* 亲液的
lyophobic *adj.* 疏液的
cluster *n.* 群；簇
anionic *adj.* 阴离子的
n. 阴离子
amphoteric *adj.* 两性的
quaternary ammonium 季铵
carboxylate *n.* 羧酸盐
monovalent *adj.* 一价的
divalent *adj.* 二价的
protonate *v.* （使）质子化
polydimethylsiloxane *n.* 聚二甲硅氧烷
phosphate *n.* 磷酸盐
poly(ethylene glycol) 聚乙二醇
polyether *n.* 聚醚
fertilizer *n.* 肥料
bactericide *n.* 杀菌剂
isoelectric *adj.* 等电位的
zwitterionic *adj.* 两性离子的
n. 两性离子

Notes

➢ The relative size of the hydrophobic and polar groups, not the absolute size of either of the two, is decisive in determining the physicochemical behaviour of a surfactant in water.

参考译文：疏水基团和极性基团的相对大小，而非两者的绝对大小，是表面活性剂在水中的物理化学行为的决定性因素。

➢ Surfactants based on sulfate or sulfonate to give a negative charge remain zwitterionic down to very low pH values due to the very low pK_a values of monoalkyl sulfuric acid and alkyl sulfonic acid, respectively.

参考译文：由于单烷基硫酸和烷基磺酸的 pK_a 值很低，基于硫酸盐或磺酸盐的表面活性剂在极低的 pH 值下仍然为两性离子。

➢ One characteristic feature of surfactants is their tendency to adsorb at interfaces. Another fundamental property of surface active agents is that unimers in solution tend to form aggregates, so-called micelles.

参考译文：表面活性剂的一个特征是它们倾向于在界面处吸附。表面活性剂的另一个基本性质是溶液中的单聚体倾向于形成聚集体，即所谓的胶束。

➢ Fatty alcohol ethoxylates can be regarded as hydrolytically stable in the pH range of 3~11. They undergo a slow oxidation in air, however, and some oxidation products, e.g. aldehydes and hydroperoxides, are more irritating to the skin than the intact surfactant.

参考译文：脂肪醇乙氧基酯在 pH 值 3~11 范围内被认为是水解稳定的。然而它们在空气中经历缓慢的氧化形成的一些氧化产物，如醛和氢过氧化物，比完整的表面活性剂更刺激皮肤。

➢ An amphoteric surfactant is one that changes from net cationic via zwitterionics to net anionic on going from low to high pH. Neither the acid nor the base site is permanently charged, i.e. the compound is only zwitterionic over a certain pH range.

参考译文：在 pH 值从低到高的过程中，两性表面活性剂的两性离子从净阳离子变成净阴离子。酸和碱都不是一直存在的，也就是说，这种化合物只有在一定的 pH 值范围内才是两性离子。

Exercises

1. Discuss the following questions

(1) What is the definition of surfactants?

(2) What is the definition of critical micelle concentration?

(3) How to divide surfactants?

(4) What are the characteristic features of surfactants?

2. Put the following into Chinese

immiscible　amphoteric　isoelectric　micellization　phosphate　lyophilic

zwitterionic　poly(ethylene glycol)　critical micelle concentration

3. Put the following into English

表面张力　亲水的　胶体的　胶束　两亲的

质子化　簇　二价的　疏液的

4. Translate the following into Chinese

(1) In a micelle, the surfactant hydrophobic group is directed towards the interior of the cluster and the polar head group is directed towards the solvent. The micelle, therefore, is a polar aggregate of high water solubility and without much surface activity. When a surfactant adsorbs from aqueous

solution at a hydrophobic surface, it normally orients its hydrophobic group towards the surface and exposes its polar group to the water. The surface has become hydrophilic and, as a result, the interfacial tension between the surface and water has been reduced.

(2) The driving force for a surfactant to adsorb at an interface is to lower the free energy of that phase boundary. The interfacial free energy per unit area represents the amount of work required to expand the interface. The term interfacial tension is often used instead of interfacial free energy per unit area. Thus, the surface tension of water is equivalent to the interfacial free energy per unit area of the boundary between water and the air above it.

UNIT 4 Fragrances and Flavors

If there is a large piece of raw meat in front of you, you may have no appetite, but it must be "really fragrant" after cooked. Thanks to our ancestors for learning to use fire, so that food can undergo various wonderful chemical reactions under the catalysis of heating to generate rich flavor substances.

The fragrances of flowers, fruits, and fauna of natural scents have long been highly popular. What makes this attraction possible? What is the chemistry of fragrances and flavors? This unit will focus on the chemical basis of fragrance attraction and present a rich tapestry of chemistry and the vibrant fragrance industry behind our emotional attraction to scents and perfumes. Perfumery examples delineating how minor changes in the structure, stereochemistry, shape, and chirality of a fragrance ingredient or a scent molecule influence its smell, pleasantness, and usefulness are highlighted. Types of flavoring substances are summarized in Table 4.3.

Fragrance and flavor attraction are intertwined through a multi-sensory communication of pheromones and neurotransmitters. Why are bees attracted to flowers? Why does a humming bird typically feed on red flowers? Besides pollination, it has been speculated that flowers developed fragrances, spicy smelling natural scents not only to attract insects but also to act as a defense mechanism to ward off plant eaters.

The attractive aroma of barbecue mainly comes from the famous Maillard reaction, lipid oxidation and thiamine degradation. Maillard reaction is also the delicious secret of many foods such as toast and braised pork. Under heating conditions, proteins or amino acids react with reducing sugars (glucose, fructose, etc.) to produce a series of substances, including melanin, which makes food brown, and flavor substances.

Researchers analyzed a large number of sulfur-containing compounds (dimethyl disulfide, dimethyl trisulfide, mercaptans, thiophenes, etc.) and oxygen- and nitrogen-containing heterocyclic compounds (furan, pyrrole, etc.). The aroma of roast duck has also been separated into as many as 90 compounds, including aldehydes, hydrocarbons, ketones, alcohols, esters, phenols, heterocycles and so on. Among them, the key to "meat flavor" is sulfur-containing compounds.

People are naturally attracted to smells that are pleasing, soothing, and calming. As a result, there has been a significant rise in the use of homecare fragrances, incense sticks, and candles. Fragrances have been known to evoke childhood memories and can touch our hearts, which has been used for beneficial purposes like aromatherapy. Fragrances also impart feeling like naturalness, cleanliness, softness, and pleasantness. As a result, they are widely used in personal hygiene and homecare products. Thanks to the technological advances in the art of organic synthesis, perfumes, once the luxury of kings and queens, have now become available for everyone to enjoy.

Table 4.3 Types of flavoring substances

Type	Description
Natural flavoring substances	These flavoring substances are obtained from plant or animal raw materials, by physical, microbiological, or enzymatic processes. They can be either used in their natural state or processed for human consumption, but cannot contain any nature-identical or artificial flavoring substances.
Nature-identical flavoring substances	These are obtained by synthesis or isolated through chemical processes, which are chemically and organoleptically identical to flavoring substances naturally present in products intended for human consumption. They cannot contain any artificial flavoring substances.
Artificial flavoring substances	These are not identified in a natural product intended for human consumption, whether or not the product is processed. These are typically produced by fractional distillation and additional chemical manipulation of naturally sourced chemicals, crude oil, or coal tar. Although they are chemically different, in sensory characteristics they are the same as natural ones.

Because of the enormous utility of fragrances, the attraction to fragrances and flavors has become a huge global business worth $26.3 billion in 2017. The top five flavors and fragrances companies control 61.5% of the total share of the market. In addition to the chemical reaction from the meat's own components (proteins, lipids, etc.), various seasonings and spices are also an important source of barbecue aroma. Heat accelerates the molecular movement of these aroma components, especially volatile substances, allowing them to spread out in the air and get into your nose all the way. So, if you smell the smell of barbecue, even if you "eat" the meat indirectly, the other smells are the same.

Flavorings focus on altering the flavors of natural food products such as meats and vegetables, or creating flavor for food products that do not have the desired flavors such as candies and other snacks. Most types of flavorings focus on scent and taste. Few commercial products exist to stimulate the trigeminal senses, since these are sharp, astringent, and typically unpleasant flavors. Three principal types of flavorings are used (Fig.4.15 and Table 4.4).

Most artificial flavors are specific and often complex mixtures of singular naturally occurring flavor compounds combined together to either imitate or enhance a natural flavor. These mixtures are formulated by flavorists to give a food product a unique flavor and to maintain flavor consistency between different product batches or after recipe changes. The list of known flavoring agents includes thousands of molecular compounds, and the flavor chemist (flavorist) can often mix these together to produce many of the common flavors. Many flavorings consist of esters, which are often described as being "sweet" or "fruity".

1. Extraction Methods of Flavor

Water distillation, water and steam distillation, steam distillation, cohobation, maceration and enfleurage are the most traditional and commonly used methods. Maceration is adaptable when oil yield from distillation is poor. Distillation methods are good for powdered almonds, rose petals and rose blossoms, whereas solvent extraction is suitable for expensive, delicate and thermally unstable materials like jasmine, tuberose, and hyacinth. Water distillation is the most favored method of production of citronella oil from plant material.

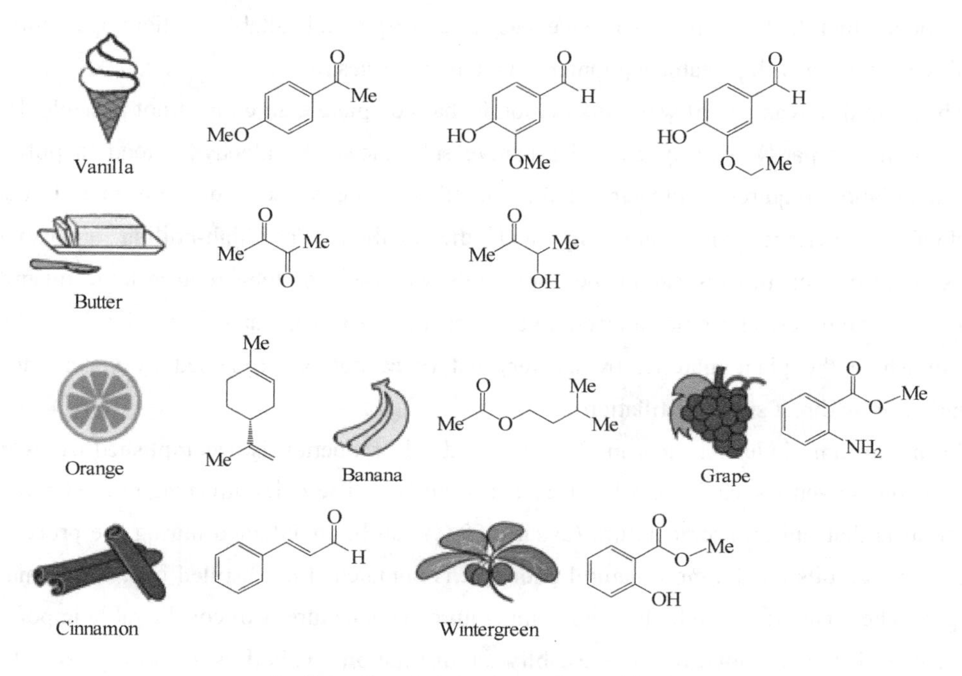

Fig.4.15 Chemical structures of several foods

Table 4.4 Chemicals and related odors

Chemical	Odor
Diacetyl, acetylpropionyl, acetoin	Buttery
Isoamyl acetate	Banana
Benzaldehyde	Bitter almond, cherry
Cinnamaldehyde	Cinnamon
Ethyl propionate	Fruity
Methyl anthranilate	Grape
Limonene	Orange
Ethyl decadienoate	Pear
Allyl hexanoate	Pineapple
Ethyl maltol	Sugar, cotton candy
Ethylvanillin	Vanilla
Methyl salicylate	Wintergreen

During water distillation, all parts of the plant charge must be kept in motion by boiling water, this is possible when the distillation material is charged loosely and remains loose in the boiling water. For this reason only, water distillation possesses one distinct advantage, i.e. that it permits the processing of finely powdered materials or plant parts that, by contact with live steam, would otherwise form lumps through which the steam cannot penetrate. Other practical advantages of water distillation are that the stills are inexpensive, easy to construct and suitable for field operation. These are still widely used with portable equipment in many countries.

The main disadvantage of water distillation is that complete extraction is not possible. Besides, certain esters are partly hydrolyzed and sensitive substances like aldehydes tend to polymerize. Water distillation requires a greater number of stills, more space and more fuel. It demands considerable experience and familiarity with the method. The high-boiling and somewhat water-soluble oil constituents cannot be completely vaporized or they require large quantities of steam. Thus, the process becomes uneconomical. For these reasons, water distillation is used only in cases in which the plant material by its very nature cannot be processed by water and steam distillation or by direct steam distillation.

In the perfume industry, most modern essential oil production is accomplished by extraction, using volatile solvents such as petroleum ether and hexane. The chief advantage of extraction over distillation is that uniform temperature (usually 50℃) can be maintained during the process. As a result, extracted oils have a more natural odor that is unmatched by distilled oils, which may have undergone chemical alteration by the high temperature. This feature is of considerable importance to the perfume industry. However, the established distillation method is of lower cost than the extraction process.

2. Synthetic Technologies Used for the Discovery of New Fragrance

Diels-Alder technology has been used extensively in the synthesis of numerous flavor and fragrance molecules that are used in consumer and fine fragrances. The odor description and structures of a few key ingredients like Lyral, aldehyde AA, isocyclocitral, Iso E super, Isoprecyclemone B, Melafleur, Camek, Oriniff, and δ-damascone are shown in Fig.4.16.

As previously stated, consumer fragrances in soaps, detergents, conditioners, and softeners employ very stringent conditions and harsh bases. Because of this, perfumers need to use fragrance ingredients that would be stable in those environments. Fragrance ingredients with aldehyde groups play a large role in the design of functional perfumes, but unfortunately, aldehydes undergo many side reactions in the basic media, such as aldol condensations and polymerizations.

It has been observed that a simple transformation of an aldehyde group into a nitrile group present in a fragrance ingredient not only enhances its base stability, but also retains its odor. Fig.4.17 highlights a select group of aromatic nitrile fragrance ingredients such as Fleuranil, Salicynalva, Khusinil, and a few acyclic nitrile fragrance ingredients such as Azuril, Peonile, Citralva, Citronalva, and Lemonalva that perform well in basic applications.

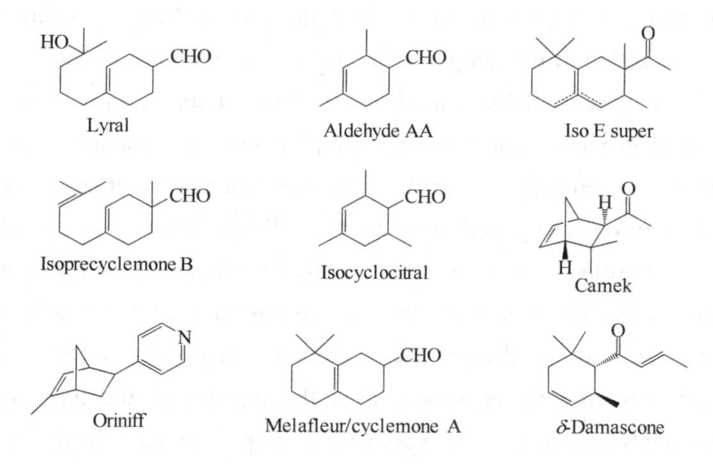

Fig.4.16 Fragrance ingredients based on Diels-Alder technology

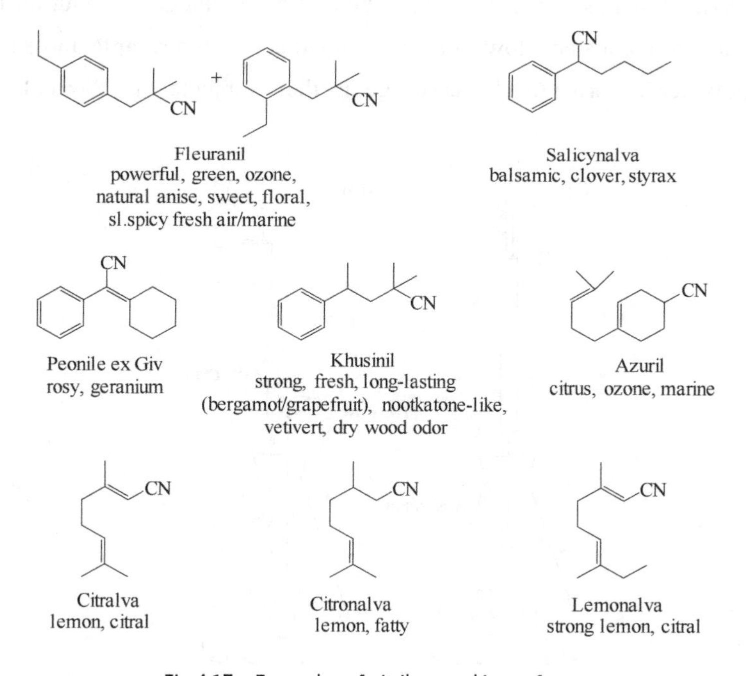

Fig.4.17 Examples of nitriles used in perfumery

3. Analysis of Fragrance and Flavor

Fragrance chemists use instrumental analysis techniques, such as gas chromatography-mass spectrometry (GC-MS), nuclear magnetic resonance (NMR), infrared spectroscopy (IR), and ultraviolet (UV), to analyze and identify various components that are present in the essential oils. GC-olfactometry is an important technique that is used to identify key components of a given essential oil, fruit, or spice. In addition, synthesis can be used to corroborate the structure of an unknown aroma chemical present in these natural fragrance ingredients.

The most commonly employed method to obtain quantitative data is gas chromatography (GC), either after solvent extraction of the sample or in combination with headspace sampling. Headspace techniques, which have successfully been used for the identification of volatiles emitted from plants, have the advantage of generally not requiring complex sample preparation, as the volatiles are directly trapped above the sample by reversible adsorption on a polymer substrate and analyzed by GC after thermal desorption. The experiments are typically carried out in a closed container. Static headspace analysis, also referred to as solid-phase microextraction (SPME), allows determination of the composition of the gas phase which is in equilibrium with the solid or liquid sample [Fig.4.18(a)]. In a typical set-up, the polymer substrate is fixed on the top of a syringe needle, which after analysis can directly be desorbed in the injector of the gas chromatography. In dynamic headspace analysis, the gas phase above the sample is continuously removed to account for the convection to which a sample is subjected when exposed to the air [Fig.4.18(b)]. This allows the evaporation kinetics of the volatiles to be monitored under non-equilibrium conditions. In a typical experiment, a constant flow of air is pumped across the sample and across a cartridge containing the polymer adsorbent. The cartridge is then subjected to thermal desorption and analyzed by GC.

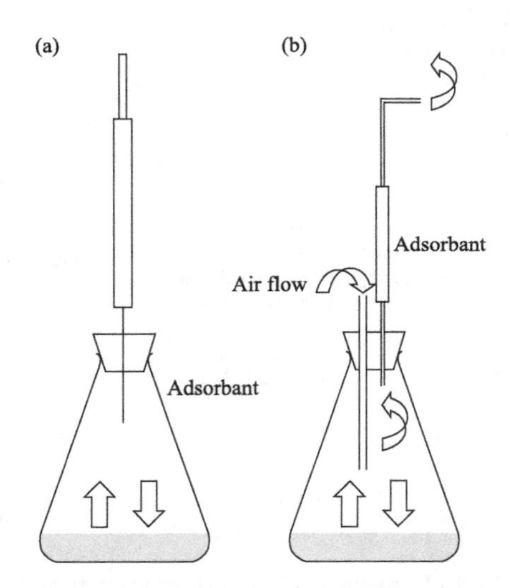

Fig.4.18 The principle of (a) static and (b) dynamic headspace sampling

4. Future of Fragrances: Natural, Organic, Healthy, and Sustainable

Going forward, more and more consumers will use fragrances not just for sensory and esthetic benefits but they will also prefer to purchase scents that are healthy, holy, and sustainable. This trend of wellness is already gaining momentum. It is no wonder that there is an increasing interest in the use of natural essential oils in perfumes and aromatherapy. Concurrently, the use of niche fragrances is growing exponentially in the industry.

With regard to holy and healthy fragrances, Asian cultures have relied on the use of holy basil, sandal, natural spices and attars of selected flowering plants for healing purposes for many years. There are numerous fragrance healing gardens in Kashmir, India dating back to the 1700s.

In moving forward toward the goal of assuring sustainable fragrances, synthetic biotechnology will be a major source of key natural flavor and fragrance ingredients. In addition to using natural ingredients, synthetic biotechnology will also be used for the production of sustainable raw materials and essential oils needed for creating attractive fragrances. This is primarily due to dwindling resources, climate changes, and a lack of abundant, cost-effective key fragrance ingredients without which modern fragrances cannot be created.

Words and Expressions

fragrance *n.* 香精；香味
flavor *n.* 香料
tapestry *n.* 挂毯；织锦
vibrant *adj.* 生气勃勃的
perfume *n.* 香水
chirality *n.* 手性
intertwined *adj.* 错综复杂的
pheromone *n.* 信息素
neurotransmitter *n.* 神经递质
incense stick 卫生香；香薰棒
aromatherapy *n.* 芳香疗法
hygiene *n.* 保健；卫生学
stimulate *v.* 刺激；促进
trigeminal *adj.* 三叉神经的
astringent *adj.* 涩的
enzymatic *adj.* 酶的；酶促
consumption *n.* 消费
imitate *v.* 模仿；效仿
recipe *n.* 食谱；秘诀
odor *n.* 气味
isoamyl acetate 乙酸异戊酯
bitter almond 苦杏仁
ethyl propionate 丙酸乙酯
ethyl decadienoate 癸二烯酸乙酯
allyl hexanoate 己酸烯丙酯
ethyl maltol 乙基麦芽酚
olfactometry *n.* 嗅觉测量法
sandal *n.* 檀香

Notes

➢ Maillard reaction is also the delicious secret of many foods such as toast and braised pork. Under heating conditions, proteins or amino acids react with reducing sugars (glucose, fructose, etc.) to produce a series of substances, including melanin, which makes food brown, and flavor substances.

参考译文：美拉德反应也是烤面包、红烧肉等诸多食物美味的秘诀。在加热条件下，蛋白质或氨基酸与还原糖类（葡萄糖、果糖等）反应生成一系列物质，包括让食物呈现褐色的黑素，以及香味物质。

➢ Heat accelerates the molecular movement of these aroma components, especially volatile substances, allowing them to spread out in the air and get into your nose all the way.

参考译文：热量会加速这些香气成分尤其是挥发性物质的分子运动，使它们在空气中扩散并进入你的鼻子内。

➢ Few commercial products exist to stimulate the trigeminal senses, since these are sharp, astringent, and typically unpleasant flavors.

参考译文：很少有商品会去刺激你的三叉神经，因为这些产品味道刺激、涩，并且通常令人感到不舒服。

➢ Other practical advantages of water distillation are that the stills are inexpensive, easy to construct and suitable for field operation.

参考译文：水蒸馏的其他实际优点包括蒸馏器价格低廉、易于建造且适合现场操作。

➢ The most commonly employed method to obtain quantitative data is gas chromatography (GC), either after solvent extraction of the sample or in combination with headspace sampling.

参考译文：气相色谱法是获得定量数据最常用的方法，无论是在溶剂提取样品后还是与顶空进样结合使用。

Exercises

1. Discuss the following questions

(1) What are three principal types of flavorings used?

(2) What is the important technique that is used to identify key components of fragrance and flavor?

(3) What is the future development trend of flavors and fragrances?

2. Put the following into Chinese

appetite	pheromone	thiamine	fructose	artificial
consistency	distillation	polymerize	steam	ingredients

3. Put the following into English

气味	醛类	零食	提取	组分
汽油	己烷	冷凝	紫外的	蒸发

4. Translate the following into Chinese

(1) Most artificial flavors are specific and often complex mixtures of singular naturally occurring flavor compounds combined together to either imitate or enhance a natural flavor. These mixtures are formulated by flavorists to give a food product a unique flavor and to maintain flavor consistency between different product batches or after recipe changes. The list of known flavoring agents includes thousands of molecular compounds, and the flavor chemist (flavorist) can often mix these together to produce many of the common flavors. Many flavorings consist of esters, which are often described as being “sweet” or “fruity”.

(2) The main disadvantage of water distillation is that complete extraction is not possible.

Besides, certain esters are partly hydrolyzed and sensitive substances like aldehydes tend to polymerize. Water distillation requires a greater number of stills, more space and more fuel. It demands considerable experience and familiarity with the method. The high-boiling and somewhat water-soluble oil constituents cannot be completely vaporized or they require large quantities of steam. Thus, the process becomes uneconomical.

UNIT 5 Chiral Drugs

1. Chirality in Nature: You Can Only Use One of Them

So far, mother nature seems to have some affinity for certain stereoisomers as well.

Let's start by looking at the building blocks of proteins, amino acids. The general structure of amino acids can be given by Fig.4.19, where X can be as simple as a hydrogen atom in the case of glycine or as complex as an indole in the case of tryptophan. Except for the case of glycine (which is prochiral anyway), all amino acids are chiral compounds since the carbon numbered 1 is a chiral center. This means that these amino acids have enantiomers. However, all amino acids that are utilized in eukaryotic genetic structures are the *l*-stereoisomers, or left-hand type, that is, these amino acids specifically have the structure shown on the right. While *d*-stereoisomers do of course exist, these amino acids never get picked by ribosomes during translation at all.

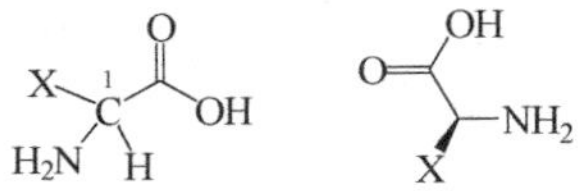

Fig.4.19 Stereoisomers in nature

On the other hand, naturally occurring sugars tend to be the *d*-stereoisomers or the right-hand type. Glucose, for example, only occurs naturally as a *d*-glucose. *l*-glucose has to be synthesized in the laboratory. While both *d*-glucose and *l*-glucose taste pretty much the same, yet our body is specific and will only bind to *d*-glucose (and other *d*-hexose). As a result, if by some slim chance you are eating *l*-glucose, it will taste sweet as usual, but it will pass right through and not be absorbed into the body. Too bad *l*-glucose is still too expensive to manufacture, so we cannot have that as a substitute sweetener yet.

This property also extends to deoxyribose sugar that is a precursor of DNA (deoxyribonucleic acid). Deoxyribose sugar naturally occurs in the *d*-form. Combining this with the absolute stereochemistry of the nitrogenous bases, the nucleosides all have the *R* configuration about C_1. This causes the double helical structure of the DNA to be right-hand in its normal state (Fig.4.20).

Another interesting "application" of chirality in nature deals with tastes and smells. Interestingly, *R*-(−)-carvone smells similar to spearmint leaves while *S*-(+)-carvone smells similar to caraway seeds despite the similar structure. This is because the olfactory receptors are chiral. As a result, they respond differently to two enantiomers. Another pair of enantiomers also exhibit this property. *l*-limonene is found in mint oil and smells like turpentine, while *d*-limonene is found in orange and, you guess it, smells like orange.

2. What is Chirality?

Chirality is derived from the Greek word χειρ (kheir) that stands for "hand". An object is said to be chiral if the object and its mirror image are non-superimposable, just like our right and left hands. Now you must be wondering what we mean by non-superimposable. When the mirror image of the object is placed over the original object and they do not overlap, as shown in Fig.4.21, then the object and its image are said to be non-superimposable.

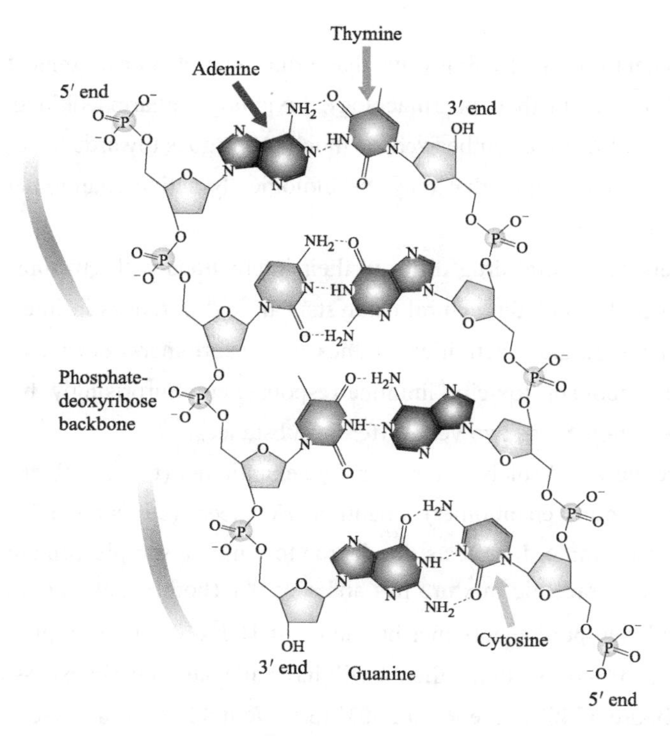

Fig.4.20 Structure of DNA

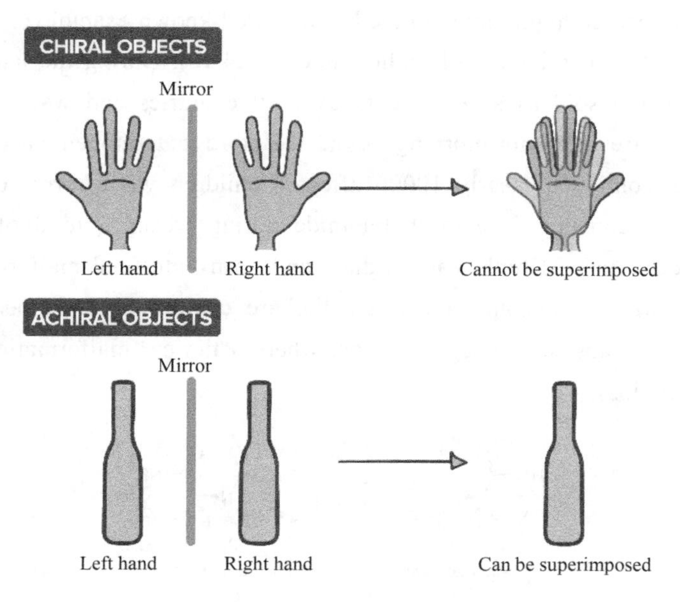

Fig.4.21 Chirality and achirality

3. Chiral Drugs

The importance of chiral drugs in the drug development space cannot be understated. In pharmaceutical industries, 56% of the drugs currently in use are chiral molecules and 88% of the last ones are marketed as racemates (or racemic mixtures), consisting of an equimolar mixture of two enantiomers.

Although the enantiomers of chiral drugs have the same chemical connectivity of atoms, they exhibit marked differences in their pharmacology, toxicology, pharmacokinetics, metabolism, etc. Therefore, when chiral drugs are synthesized, as much effort goes towards the rigorous separation of the two enantiomers. This ensures that only the biologically active enantiomers are present in the final drug preparation.

The enantiomers of a chiral drug differ in their interactions with enzymes, proteins, receptors and other chiral molecules including chiral catalysts. These differences in interactions, in turn, lead to differences in the biological activities of the two enantiomers, such as their pharmacology, pharmacokinetics, metabolism, toxicity, immune response, etc. Surprisingly, biological systems can recognize the two enantiomers as two very different substances.

Some drugs are marketed solely as a pure single enantiomer (that is, the drug preparation has no contamination with the other enantiomer). Enantiomeric excess (*ee*) is a measurement of the degree of purity of any chiral sample. It reflects the degree to which a sample contains one enantiomer in excess over the other. A racemic mixture has an *ee* of 0% (both enantiomers are present in a 1∶1 ratio), while a completely pure enantiomer has an *ee* of 100%. As an example, if a sample contains 70% of *R* isomer and 30% of *S* isomer, then it will have an enantiomeric excess of 40%. This can be rationalized as a mixture of 40% pure *R* with 60% (30% *R* and 30% *S*) of a racemic mixture.

Enantiomer ratio is extremely important because while one enantiomer is beneficial to the body, the other enantiomer can be highly toxic to the body. A well-known example of enantiomer-related toxicity is the *R*- and *S*-enantiomers of thalidomide (Fig.4.22). During the late 1950s and early 1960s, thalidomide was sold as a sedative in over 40 countries and was often prescribed to pregnant women as a treatment for morning sickness. Before its teratogenic activity came to light and its use was discontinued, nearly 10000 affected children were born from women taking thalidomide during pregnancy. Use of thalidomide during weeks 3 to 8 of gestation causes multiple birth defects such as limb, ear, cardiac, and gastrointestinal malformations. The limb malformations, known as phocomelia and amelia, are characterized, respectively, by severe shortening or complete absence of legs or arms, whereas the ear malformations lead to anotia, microtia, and hearing loss.

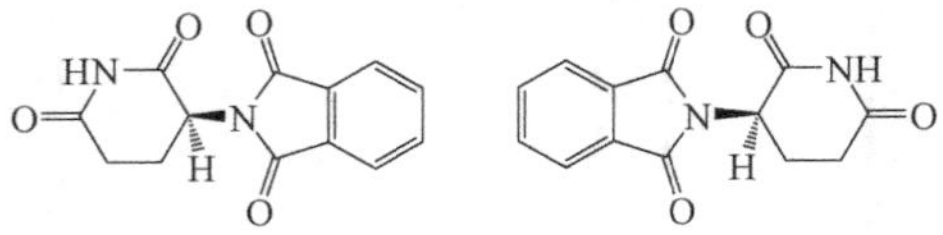

R-thalidomide　　*S*-thalidomide

Fig.4.22 Molecular formula of thalidomide

4. Why Do Enantiomers Have Different Biological Activities?

Recognition of chiral drugs by specific drug receptors is explained by a three-point interaction of the drug with the receptor site, as proposed by Easson and Stedman. The difference between the interaction of the two enantiomers of a chiral drug with its receptor is illustrated below (Fig.4.23).

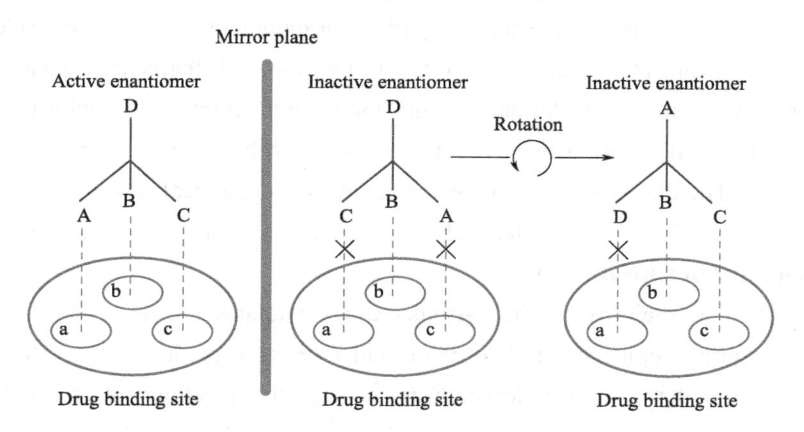

Fig.4.23 Three-point interaction of the drug with the receptor site

Easson-Stedman's illustration of hypothetical interaction between the two enantiomers of a racemic drug with a receptor at the drug binding sites: The three substituents A, B, C of the active enantiomer (left) can interact with three binding sites a, b, c of a receptor by forming three contacts Aa, Bb and Cc, whereas the inactive enantiomer (right) cannot because the contact is insufficient.

In this case, one enantiomer is biologically active while the other enantiomer is not. The substituents of the active enantiomer drug labelled A, B, and C must interact with the corresponding regions of the binding sites labelled a, b, and c of the receptor in order to have a proper alignment Aa, Bb, Cc. In this case, this fitting interaction produces an active biological effect. In contrast, the inactive enantiomer cannot bind in the same way with its receptor. Thus, there is no active response. The attachment of an enantiomer to the chiral receptor is analogous to a hand fitting into a glove. Indeed, a right hand can only fit into a right-hand glove. Similarly, only a particular enantiomer that has a complementary shape to a receptor site can fit into the receptor site. The other enantiomer will not fit, like a right hand will never fit into a left glove.

5. Separation Technique of Chiral Drugs

Chiral drugs are separated both in pharmaceutical industry and for clinical purposes. It is a procedure in which the isomers of a racemic mixture are separated. Many drugs during their synthesis from pure enantiomeric reagents, often lead to the formation of a racemic mixture. Pharmaceutical industries follow two approaches: the classical method and the modern technology.

The classical approach uses the formation of a diastereomeric salt followed by its resolution. It involves an acid-base reaction between a racemic drug and a pure single enantiomer. The salt formed has different physical and chemical properties which are separated by crystallization or filtration, in which the salt is treated either with an acid or a base to yield the pure enantiomer. Another method of separation involves achiral liquid chromatography which is used to separate asparagine, methyl-

L-dopa, and glutamic acid. Enzymatic or kinetic resolution uses microorganisms like yeasts, moulds, or bacteria that destroy one of the enantiomeric forms.

In modern technology, high-performance liquid chromatography HPLC\left (HPLC\right) is used to separate the enantiomers. This method is classified further as direct and indirect. In the direct method, a chiral selector is used in the stationary phase or mobile phase. The chiral selector uses the lock-and-key mechanism. However, several other factors like pH, temperature, length of column, etc., play an important role for resolution. This method is rarely employed in industries because it is too costly and has low efficiency. The indirect method starts with a pure single enantiomer and forms two diastereomers. The diastereomers can be separated by the classical reverse phase column. It is rarely used in industries but is used in the analysis of biological samples due to its high sensitivity.

6. Development of Chiral Drugs

Until recently, the majority of single-isomer drugs available are those derived from natural sources (e.g., morphine, epinephrine, hyoscine), and racemates predominated. There is no clear evidence of a trend in the pharmaceutical industry towards the development of chiral drugs, either denovo or by deriving them from racemates marketed previously (i.e., chiral switching). Several factors have influenced this trend, which has occurred independently and in parallel with a quest in the industry as a whole to develop more potent, selective and specific drugs. Chiral switching is a small component of this trend and relies on the existence of a racemate in the first place. The process is equivalent to developing a new active substance and requires a new application, but data on the racemate may be used as appropriate, together with bridging studies. There is, however, limited potential in the market for the degree of therapeutic benefit obtained to justify the degree of investment.

The real benefit of chiral technology lies in its application in the search for novel chemical entities. Regarding enantiomers as chemically distinct entities at an early stage in the research and development process is a valuable aid in the understanding of drug mechanisms. The key targets of selectivity and specificity can be pursued in an effort to improve drug efficacy while minimizing toxicity. Rational drug design (based on increased understanding of biological receptor systems) combined with chiral technology allows a "chiral template" to be developed.

Coinciding with this has been the awakening of the drug regulatory authorities to the different pharmacological and toxicological profiles of enantiomers. The FDA policy statement for the development of stereoisomeric drugs, issued in 1992, made it more difficult to obtain approval for racemates. This statement made it clear that approval could not be granted for a drug containing more than one isomer unless the pharmacokinetic and pharmacodynamic properties of each could be described and, more importantly, justified. In addition, the FDA offers a shortened approval process for the enantiomeric versions of approved drugs, with the promise of patent protection.

Drug regulatory authorities in other countries have followed this lead. Additional isomers in a compound are no longer considered as "silent passengers" but as potential contaminants (so-called isomeric ballast). While it is unlikely that many racemates will be approved in the future as a result of this, there are still situations in which their production is justified. These include situations in which ①the enantiomers are configurationally unstable in vitro or undergo racemization in vivo; ② the enantiomers have similar pharmacokinetic, pharmacodynamic and toxicological properties; and ③it is not technically feasible to separate the enantiomers in sufficient quantity and quality.

Situations also exist in which an enantiomeric ratio other than unity may be justified if the ratio is expected to improve the therapeutic profile, as there is no reason to expect the optimum eudismic ratio to be necessarily 1:1 (i.e., the dose-response curves would not usually be expected to be congruent).

Words and Expressions

protein *n.* 蛋白质
tryptophan *n.* 色氨酸
ribosome *n.* 核糖体
glucose *n.* 葡萄糖
helical *adj.* 螺旋形的
superimpose *v.* 叠加
pharmacology *n.* 药物学；药理学
separation *n.* 分离；分开
thalidomide *n.* 沙利度胺
receptor *n.* 受体
illustration *n.* 说明；插图；例证；图解
racemic *adj.* 外消旋的
biologically *adv.* 生物学上；生物学地
attachment *n.* 连接
complementary *adj.* 补足的；互补的
microorganism *n.* 微生物；微小动植物
fungal *adj.* 真菌的
diastereomer *n.* 非对映异构体
potent *adj.* 有效的
optimum *adj.* 最适宜的
n. 最佳效果；最适宜条件

Notes

➢ Chirality is derived from the Greek word χειρ (kheir) that stands for "hand". An object is said to be chiral if the object and its mirror image are non-superimposable, just like our right and left hands.

参考译文：手性源自希腊语单词χειρ(kheir)，代表"手"。如果一个物体和它的镜像是不可叠加的，就像我们的右手和左手一样，那么这个物体具有手性。

➢ The enantiomers of a chiral drug differ in their interactions with enzymes, proteins, receptors and other chiral molecules including chiral catalysts.

参考译文：手性药物的对映异构体与酶、蛋白质、受体和其他手性分子(包括手性催化剂)的相互作用也不同。

➢ Enantiomer ratio is extremely important because while one enantiomer is beneficial to the body, the other enantiomer can be highly toxic to the body.

参考译文：对映异构体比例非常重要，因为虽然一种对映异构体对身体有益，但另一种对映异构体可能对身体有剧毒。

➢ Chiral drugs are separated both in pharmaceutical industry and for clinical purposes. It is a procedure in which the isomers of a racemic mixture are separated. Many drugs during their synthesis from pure enantiomeric reagents, often lead to the formation of a racemic mixture.

参考译文：手性药物在制药工业和临床上都是分开的。这是一个分离外消旋异构体混合物的过程。许多药物在由纯对映体试剂合成的过程中，经常导致外消旋混合物的形成。

➢ The key targets of selectivity and specificity can be pursued in an effort to improve drug efficacy while minimizing toxicity.

参考译文：追求选择性和特异性的关键目标是提高药物疗效，同时最大限度地减小毒性。

Exercises

1. Discuss the following questions

(1) Why is the double helical structure of the DNA right-hand in its normal state?

(2) What kind of object can be said to be chiral?

(3) How to explain the recognition of chiral drugs by specific drug receptors?

(4) What are two approaches pharmaceutical industries follow to separate chiral drugs?

2. Put the following into Chinese

chirality tryptophan ribosome enantiomer thalidomide

illustration racemic attachment isomer optimum

3. Put the following into English

蛋白质 葡萄糖 不可叠加的 药理学 分离

活性 镇静剂 受体 互补的 非对映体

4. Translate the following into Chinese

(1) The real benefit of chiral technology lies in its application in the search for novel chemical entities. Regarding enantiomers as chemically distinct entities at an early stage in the research and development process is a valuable aid in the understanding of drug mechanisms. The key targets of selectivity and specificity can be pursued in an effort to improve drug efficacy while minimizing toxicity. Rational drug design (based on increased understanding of biological receptor systems) combined with chiral technology allows a "chiral template" to be developed.

(2) One enantiomer is biologically active while the other enantiomer is not. The substituents of the active enantiomer drug labelled A, B, and C must interact with the corresponding regions of the binding sites labelled a, b, and c of the receptor in order to have a proper alignment Aa, Bb, Cc. In this case, this fitting interaction produces an active biological effect. In contrast, the inactive enantiomer cannot bind in the same way with its receptor. Thus, there is no active response.

PART 5

Advanced New Techniques in Fine Chemicals

UNIT 1 Recent Progresses in Fluorescent Dyes and Chemosensors

Fluorescence is a process that involves photon absorption by a fluorophore giving an excited state and relaxation of the excited-state by emission of another photon. Fluorophore is a fluorescent moiety that can consist of disparate chemical structures, including small molecules, proteins, and semiconductor beads. Fluorescent chemosensors are defined as chemical systems that transform stimuli into fluorescence. A typical chemosensor contains a recognition site (the receptor), which is connected to the signal reporter, such as a fluorophore or chromophore. As an efficient approach to convert the act of detection into a fluorescent signal already on the molecular level, fluorescent chemosensors provide accurate quantitative detection toward analytes.

1. Luminescence

Luminescence is the emission of light from any substance, and occurs from electronically excited states. "Luminescence" comes from the Latin (lumen = light), and it was first introduced as "luminescenz" by the physicist and science historian Eilhardt Wiedemann in 1888, to describe "all those phenomena of light which are not solely conditioned by the rise in temperature", as opposed to incandescence. Luminescence is cold light whereas incandescence is hot light. There are various types of luminescence (Fig.5.1), in which fluorescence and phosphorescence belong to photoluminescence.

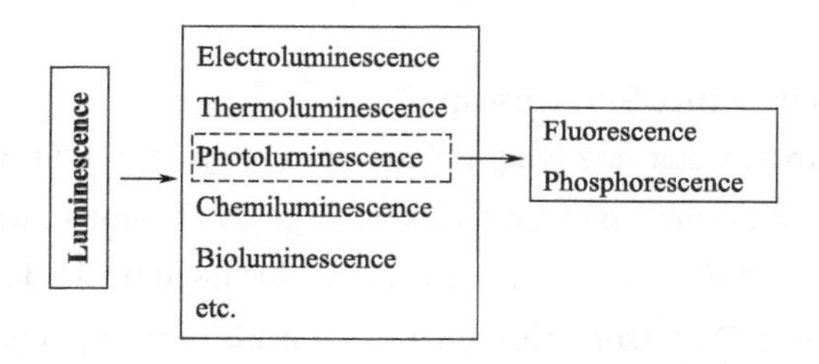

Fig.5.1 Position of fluorescence and phosphorescence in the frame of luminescence

2. Brief History of Fluorescence

The phenomena of "fluorescence" was first reported by a Spanish physician, Nicolas Monardes, in 1565. He described the wonderful peculiar blue colour of an infusion of a wood called Lignum Nephriticum. However, it was until 1852, "fluorescence" was coined as a professional term by Sir George Gabriel Stokes, a physicist and professor of mathematics at Cambridge.

Stokes reinvestigated the phenomena and published a famous paper entitled "On the refrangibility of light" in 1852. He demonstrated that the phenomenon was an emission of light following absorption of light. It is worth describing one of Stokes' experiments, which is spectacular and remarkable for its simplicity. Stokes formed the solar spectrum by means of a prism. When he moved a tube filled with a solution of quinine sulfate through the visible part of the spectrum, nothing happened: the solution simply remained transparent. But beyond the violet portion of the spectrum, i.e. in the non-visible zone corresponding to ultraviolet radiations, the solution glowed with a blue light. Stokes wrote: "It was certainly a curious sight to see the tube instantaneously light up when plunged into the invisible rays, it was literally darkness visible." This experiment provided compelling evidence that there was absorption of light followed by emission of light. Stokes stated that the emitted light is always of longer wavelength than the excited light. This statement becomes later Stokes' law.

In this first paper, Stokes called the observed phenomenon dispersive reflexion, but in a footnote, he wrote "I confess I do not like this term. I am almost inclined to coin a word, and call the appearance fluorescence, from fluorspar, as the analogous term opalescence is derived from the name of a mineral." Most of the varieties of fluorspar or fluorspath [minerals containing calcium fluoride (fluorite)] indeed exhibit the property described above. In his second paper, Stokes definitely resolved to use the word fluorescence (Fig.5.2).

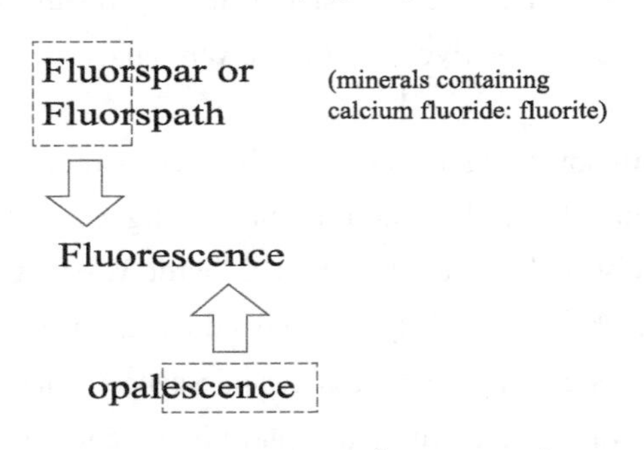

Fig.5.2 Origination of "fluorescence"

3. Principles of Fluorescence Spectroscopy

Fluorescence is a process that involving photon absorption by a fluorophore giving an excited state and relaxation of the excited-state by emission of another photon. The processes that occur between the absorption and emission of light are usually illustrated by the Jablonski diagrams. These diagrams are named after Professor Alexander Jablonski, who is regarded as the father of fluorescence spectroscopy because of his many accomplishments. A typical Jablonski diagram is shown in Fig.5.3. The singlet ground, first, and second electronic states are depicted by S_0, S_1, and S_2, respectively. At each of these electronic energy levels, the fluorophores can exist in a number of vibrational energy levels, depicted by 0, 1, 2, etc.

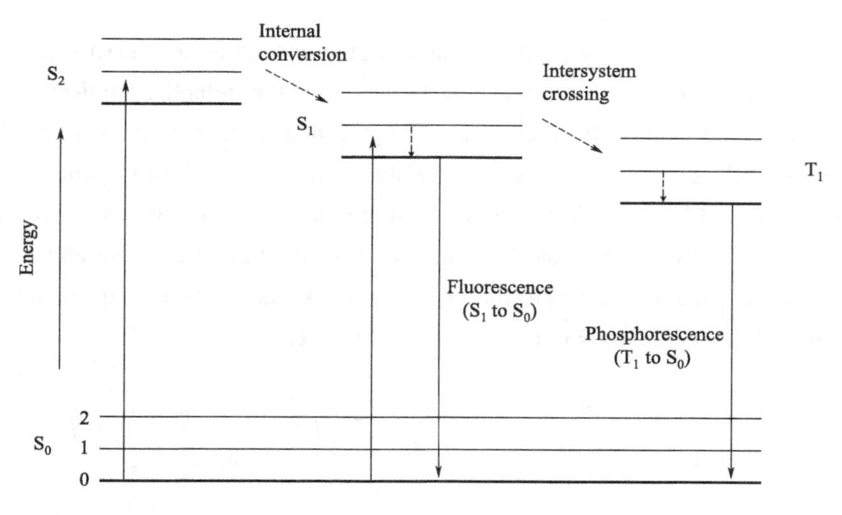

Fig.5.3 Jablonski diagram

Following light absorption, several processes usually occur. A fluorophore is usually excited to some higher vibrational level of either S_1 or S_2. With a few rare exceptions, molecules in condensed phases rapidly relax to the lowest vibrational level of S_1. This process is called internal conversion and generally occurs within 10^{-12}s or less. Since fluorescence lifetimes are typically near 10^{-8}s, internal conversion is generally complete prior to emission. Hence, fluorescence emission generally results from a thermally equilibrated excited state, that is, the lowest energy vibrational state of S_1.

Molecules in the S_1 state can also undergo a spin conversion to the first triplet state T_1. Emission from T_1 is termed phosphorescence, and is generally shifted to longer wavelengths (lower energy) relative to the fluorescence. Conversion of S_1 to T_1 is called intersystem crossing. Transition from T_1 to the singlet ground state is forbidden, and as a result the rate constants for triplet emission are several orders of magnitude smaller than those for fluorescence. Molecules containing heavy atoms such as bromine and iodine are frequently phosphorescent. The heavy atoms facilitate intersystem crossing and thus enhance phosphorescence quantum yields.

4. Common Fluorescent Dyes

There are many kinds of fluorescent dyes, among which rhodamine and fluorescein are widely used two kinds of fluorescent dyes. In addition, common fluorescent dyes include BODIPY, coumarins, cyanine dyes, etc (Fig.5.4).

5. Recent Progress in Fluorescent Dyes and Chemosensors

Currently, cytoreductive surgery followed by combination chemotherapy is regarded as the most effective treatment. The degree of cytoreduction, in which minimal residual disease is defined as tumour deposits <1cm, is one of the few prognostic factors that can be actively influenced by the surgeon. Tumour-specific intraoperative fluorescence imaging may improve staging and debulking efforts in cytoreductive surgery strategy and thereby improve prognosis. The overexpression of folate receptor-α (FR-α) in 90%~95% of epithelial ovarian cancers prompts the investigation of intraoperative tumour-specific fluorescence imaging in ovarian cancer surgery using an FR-α-targeted fluorescent agent. In patients with ovarian cancer, intraoperative tumour-specific fluorescence imaging with an FR-α-targeted fluorescent agent showcases the potential applications in patients with ovarian cancer for improved intraoperative staging and more radical cytoreductive surgery. Thus, Vasilis Ntziachristos et al. conjugated folate to fluorescein isothiocyanate (folate-FITC, Fig.5.5) for targeting FR-α together with a real-time multispectral intraoperative fluorescence imaging system. And they reported on the results of first-in-human use of intraoperative tumour-specific fluorescence imaging for real-time surgical visualization of tumour tissue in patients, undergoing an exploratory laparotomy for suspected ovarian cancer.

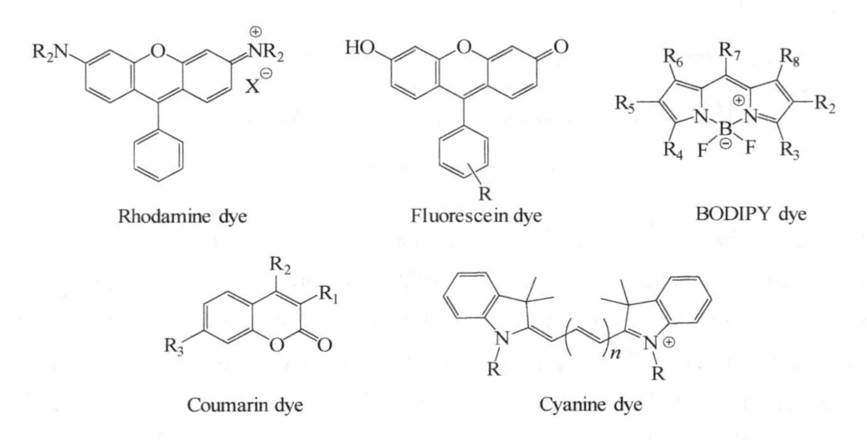

Fig.5.4 Chemical structures of common fluorescent dyes

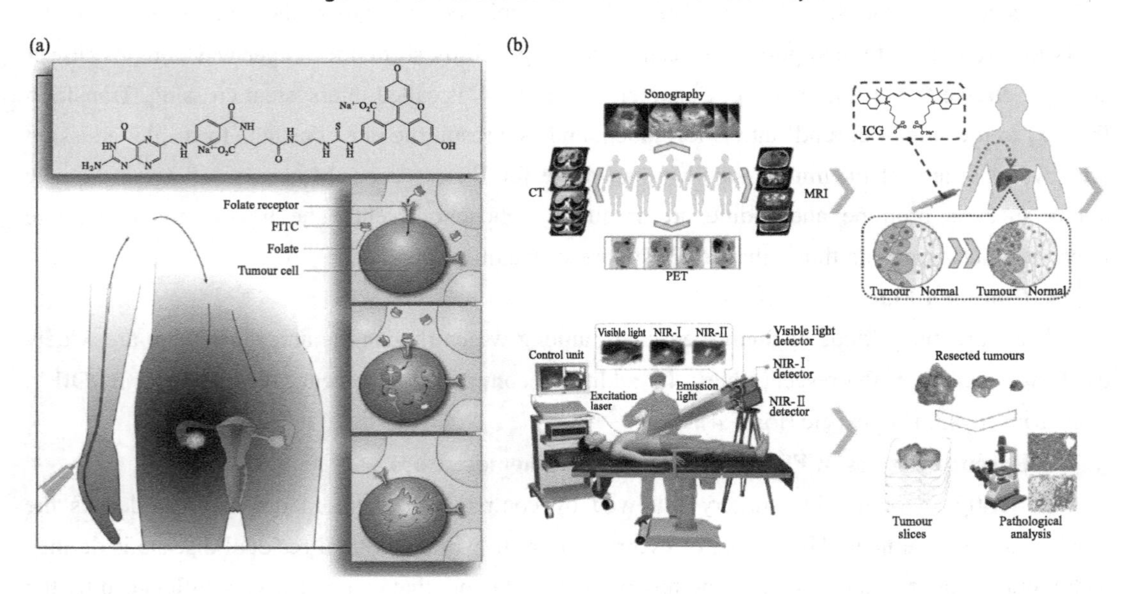

Fig.5.5 (a) Schematic presentation of the targeting of ovarian cancer; (b) Description of the study plan and the visible and NIR-I/II multispectral imaging instrument for clinical applications

The second near-infrared wavelength window (NIR-II, 1000~1700nm) enables fluorescence imaging of tissue with enhanced contrast at depths of millimetres and at micrometre-scale resolution.

However, the lack of clinically viable NIR-Ⅱ equipment has hindered the clinical translation of NIR-Ⅱ imaging. Jie Tian et al. described an optical-imaging instrument that integrates a visible multispectral imaging system with the detection of NIR-Ⅱ and NIR-I(700~900nm in wavelength) fluorescence (by using the dye indocyanine green) for aiding the fluorescence-guided surgical resection of primary and metastatic liver tumours in 23 patients [Fig.5.5(b)]. They found that, compared with NIR-I imaging, intraoperative NIR-Ⅱ imaging provided a higher tumour-detection sensitivity (100% versus 90.6%; with 95% confidence intervals of 89.1%~100% and 75.0%~98.0%, respectively), a higher tumour-to-normal-liver-tissue signal ratio (5.33 versus 1.45) and an enhanced tumour-detection rate (56.41% versus 46.15%). It can infer that combining the NIR-I/Ⅱ spectral windows and suitable fluorescence probes might improve image-guided surgery in the clinic.

Developing NIR-Ⅱ emission dye with high quantum yield and tumour-targeting ability and relative equipment can shift the paradigm of surgical oncologic imaging, thus offering the unique opportunity to intraoperatively detect and quantify tumour growth and intra-abdominal spread.

6. Development of High-performance Fluorescence Dyes for Bioimaging

High-performance fluorescence dyes based on excellent basic building block is the precondition for designing chemosensors. However, traditional fluorescence chromophores always suffer from aggregation-caused quenching (ACQ) and short-wavelength, which limits their application in in vivo fluorescent tracking of physiological processes with high-fidelity in disease diagnosis and biological research. In 2013, Zhu's group developed a high-performance building block quinoline-malononitrile (QM) with aggregation-induced emission (AIE) by replacing the oxygen atom of dicyanomethylene-4H-pyran (DCM) with *N*-ethyl group (Fig.5.6). The resulting quinoline-malononitrile (QM) derivative (ED) displayed a strong red emission, whereas the severe ACQ effect of DCM-based BD was observed in the solid state.

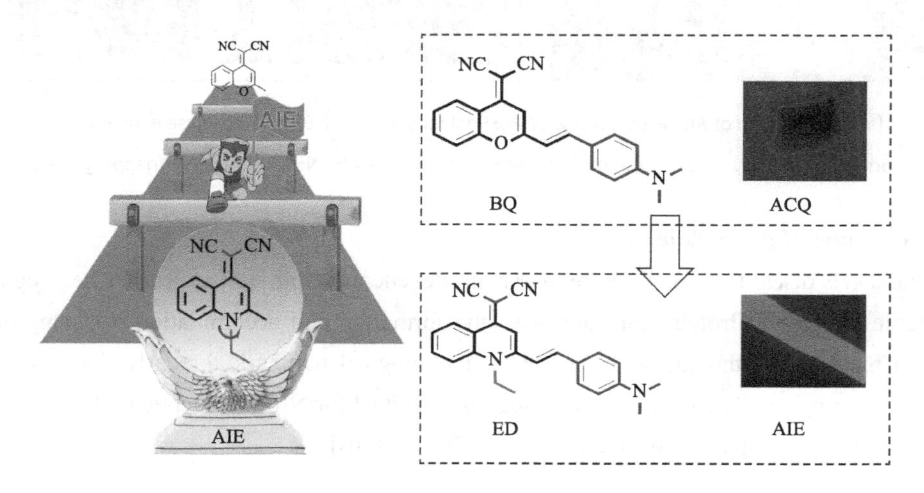

Fig.5.6 Design of high-performance AIE building block QM

Based on the building block QM, Zhu's group developed a lot of water bio-chemosensors for detection of some disease related bio-species.

(1) Sensing of Enzymes (*β*-galactosidase)

β-galactosidase (*β*-gal) is an important biomarker for cell senescence and primary ovarian cancers. Thus, an enzyme-activatable AIE-active fluorescent probe (QM-*β*-gal) (Fig.5.7) is synthesized, in which a hydrophobic AIE fluorophore, QM-OH, is utilized as a signal reporter and a hydrophilic galactose moiety as the *β*-gal-triggered unit. Due to the hydrophilic *β*-gal, QM-*β*-gal disperses well in aqueous media with non-emissive. After being activated by cellular endogenous *β*-gal, QM-*β*-gal can specifically release AIEgen QM-OH. Simultaneously, strongly emissive nanoparticles are generated in situ, providing a method for on-site sensing of endogenous *β*-gal activity in living cells. With its outstanding intracellular accumulation, the applicability of QM-*β*-gal is demonstrated for long-term tracking of *β*-gal-overexpressing SKOV-3 cells with high-resolution (Fig.5.7).

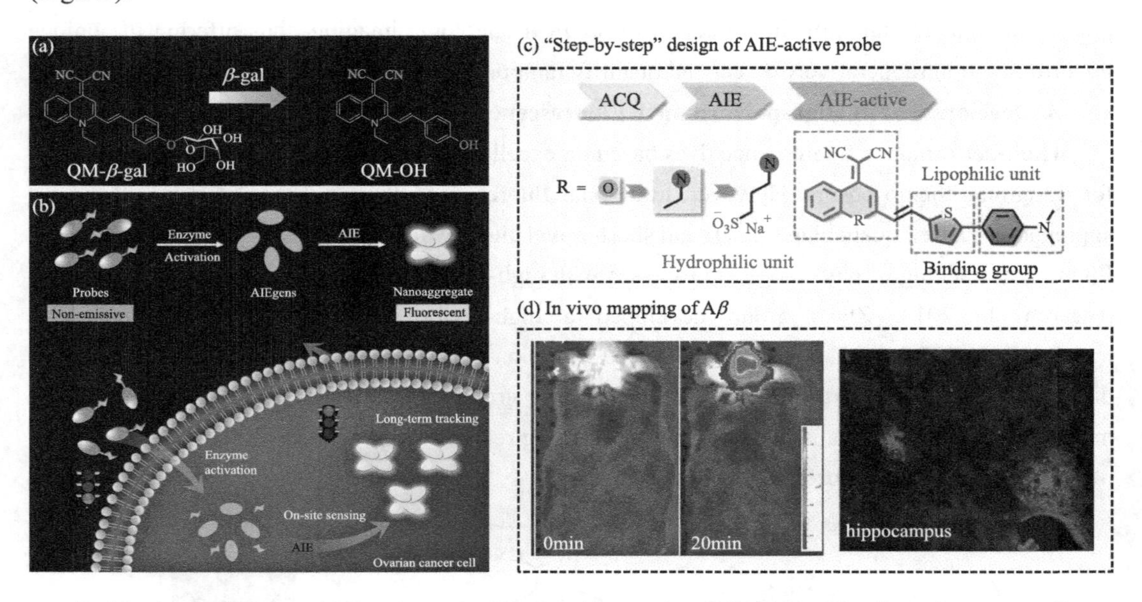

Fig.5.7 (a and b) Design of AIE-active probe QM-*β*-gal for sensing of *β*-galactosidase in ovarian cancer cell; (c and d) "Step-by-step" design of NIR AIE-active probe QM-FN-SO_3 for in vivo mapping of A*β*.

(2) Mapping of *β*-Amyloid

Alzheimer's disease (AD), a progressive neurodegenerative brain disorder, has been considered an incurable condition. Protein fibrillogenesis, the formation and accumulation of *β*-amyloid (A*β*) plaques in the brain is thought to be a critical pathological hallmark for early diagnosis of AD. Hence, the "step-by-step" rational design strategy is described for NIR AIE-active probes QM-FN-SO_3 for ultra-sensitivity mapping of A*β* plaques [Fig.5.7(c) and (d)].

Words and Expressions

luminescence *n.* 发光
fluorophore *n.* 荧光团
phosphorescence *n.* 磷光；磷光现象
chemosensor *n.* 化学传感器
aggregation-caused quenching 聚集导致淬灭
aggregation-induced emission 聚集诱导发光
hydrophobic *adj.* 疏水的
hydrophilic *adj.* 亲水的
rational design 合理的设计
β-amyloid plaque *β*-淀粉样蛋白斑块
Alzheimer's disease 阿尔茨海默病
fibrillogenesis *n.* 原纤维生成
in vivo 在活的有机体内
blood brain barrier 血脑屏障
backbone *n.* 骨架；聚合分子主链
plateau *n.* 稳定水平

Notes

➢ Fluorescence is a process that involving photon absorption by a fluorophore giving an excited state and relaxation of the excited-state by emission of another photon.

参考译文：荧光是荧光团吸收光子产生激发态，并通过发射另一个光子而使激发态弛豫的过程。

➢ The singlet ground, first, and second electronic states are depicted by S_0, S_1, and S_2, respectively. At each of these electronic energy levels, the fluorophores can exist in a number of vibrational energy levels, depicted by 0, 1, 2, etc.

参考译文：基态、第一电子态和第二电子激发态分别用 S_0、S_1 和 S_2 表示。在每个电子能级上，荧光团可以存在于许多振动能级中，用 0、1、2 等表示。

➢ The second near-infrared wavelength window (NIR-Ⅱ, 1000~1700nm) enables fluorescence imaging of tissue with enhanced contrast at depths of millimetres and at micrometre-scale resolution.

参考译文：近红外二区窗口（NIR-Ⅱ, 1000~1700nm）使组织荧光成像在毫米深度和微米尺度分辨率下具有增强的对比度。

➢ High-performance fluorescence dyes based on excellent basic building block is the precondition for designing chemosensors.

参考译文：基于优异母体结构构建的高性能荧光染料是设计化学传感器的前提。

➢ In 2013, Zhu's group developed a high-performance building block quinoline-malononitrile (QM) with aggregation-induced emission by replacing the oxygen atom of dicyanomethylene-4H-pyran (DCM) with *N*-ethyl group.

参考译文：2013 年，Zhu 的团队通过将苯并吡喃腈（DCM）的氧原子替换为氮-乙基，开发了高性能的具有聚集诱导发光性质的母体结构喹啉腈（QM）。

Exercises

1. Discuss the following questions

(1) What are the differences between fluorescence and phosphorescence?

(2) What kind of functional groups will enhance the process of intersystem crossing?

(3) What is the advantage of QM-based chromophore?

(4) What are the wavelengths of NIR-Ⅰ and NIR-Ⅱ?

2. Translate the following into Chinese

(1) "Luminescence" comes from the Latin (lumen = light), and it was first introduced as "luminescenz" by the physicist and science historian Eilhardt Wiedemann in 1888, to describe "all those phenomena of light which are not solely conditioned by the rise in temperature", as opposed to incandescence. Luminescence is cold light whereas incandescence is hot light.

(2) Traditional fluorescence chromophores always suffer from aggregation-caused quenching (ACQ) and short-wavelength, which limit their application in in vivo fluorescent tracking of physiological processes with high-fidelity in disease diagnosis and biological research.

(3) Alzheimer's disease (AD), a progressive neurodegenerative brain disorder, has been considered an incurable condition.

3. Translate the following into English

(1) β-半乳糖苷酶（β-gal）是细胞衰老和原发性卵巢癌的重要生物标志物。

(2) 大脑中 β-淀粉样蛋白（Aβ）斑块的形成和积累被认为是阿尔茨海默病（AD）早期诊断的重要病理标志。

(3) 从三线态 T_1 到单线态基态的转变是被禁止的，因此三线态发射的速率常数比荧光发射的速率常数小几个数量级。

(4) 含有重原子的分子，如溴和碘，容易发出磷光，因为重原子有利于系统间的交叉，从而提高了磷光量子产率。

UNIT 2 Organic Photochromic Molecules

1. Photochromism

The word "photochromism" (or "photochromic") stems from the Greek words, which means light and colour, respectively. The historical reference of photochromism dates back to ancient times and the era of the Alexander the Great (356—323BC). As the king of Macedonia, he got into a vast world conquest. Macedonian head warriors were equipped with photochromic bracelets exhibiting a colour change when exposed to sunlight. Such colour change was used by all warriors to indicate the right moment to begin the fight.

Photochromism is a reversible transformation of a chemical species induced in one or both directions by absorption of electromagnetic radiation between two forms, A and B, having different absorption spectra (Fig.5.8). The thermodynamically stable form A is transformed by irradiation into form B. The back reaction can occur thermally (photochromism of type T) or photochemically (photochromism of type P). Photochromic substances are widely present in glass lenses, initially clear, which turn dark under sunshine, so that eyes can be protected from harmful high light, and they are also present in trendy cosmetics and clothes. In research, photochromic materials also play an important role in data storage and transmitting, anti-counterfeiting, super-resolution imaging and molecular machine.

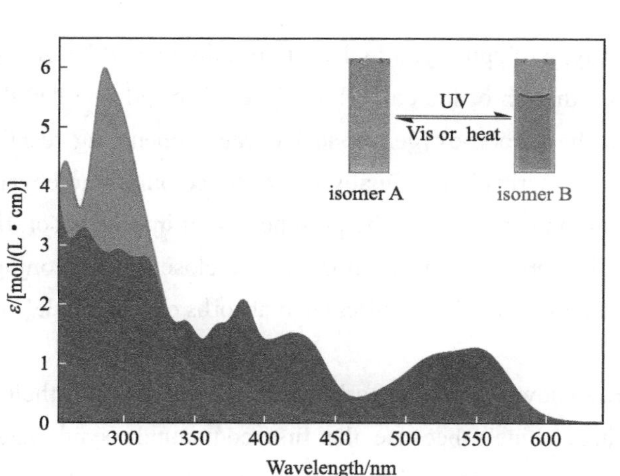

Fig.5.8 UV-Vis absorption spectra and colour changes of two isomers during photochromic process

2. Typical Photochromic Molecules

(1) Azobenzene

Azobenzene derivatives represent the main family of photochromic molecules based on *trans-cis* reaction (**1a** and **1b** in Fig.5.9). The colour of azobenzene is generally yellow and its substitution yields a bathochromic shift, providing an orange or red colour. Changes in the absorption spectra between the two forms are generally not very pronounced due to a small difference in electron delocalization between the two isomers, and the colour change is invisible by the naked eyes. In contrast, the photochromic reaction induces a very significant change of the free volume of the molecule. Therefore, these molecules are extensively used to induce mass rotation or migration in materials, leading to dichroic and birefringent media, surface relief patterning or mechanical effects.

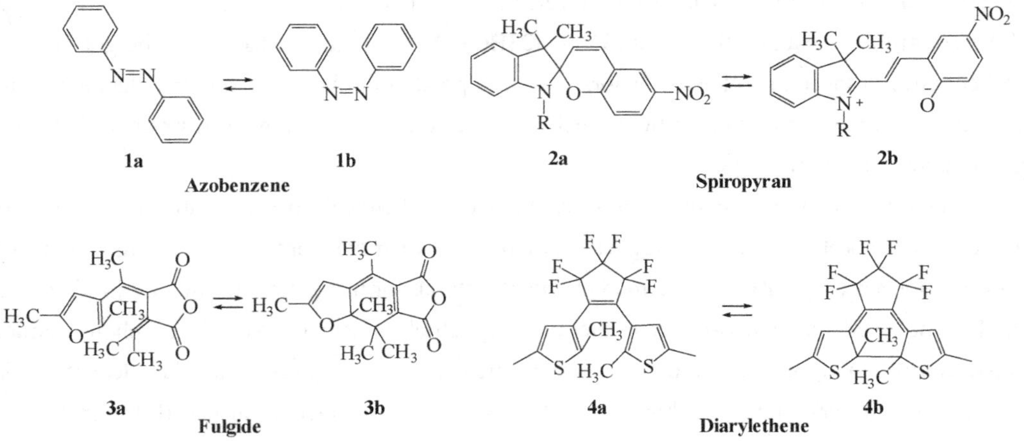

Fig.5.9 Typical examples of photochromic molecules

(2) Spiropyran

The most common type of spiropyran is based on indoline and benzopyran (Fig.5.9). For such compounds, UV light irradiation of the colourless closed form (also called the spiro or the N form, **2a** in Fig.5.9) leads to the carbon-oxygen bond cleavage (open ring reaction) of the pyran ring, followed by a *cis-trans* isomerization to finally reach the coloured merocyanine (MC) form (**2b** in Fig.5.9). This feature can be connected to the presence of an intense absorption band in the visible, and the colour of the MC form. On the contrary, in the closed form, conjugation is broken at the spiro carbon atom and the molecule is colourless and absorbs only in the UV region.

(3) Fulgide

Fulgides have been known since the early twentieth century and their name comes from the Latin "fulgere" meaning "shine", because the first compounds synthesized by Stobbe in 1905 exhibited a bright character when crystallized. Similar to spiropyrans, photochromism in fulgides is based on a cyclization reaction. The main difference with the case of spiropyrans is that the open form is colourless and the cyclized form is the coloured species (**3a** and **3b** in Fig.5.9).

(4) Diarylethene

Chemical structure of diarylethenes (DAEs) consists of two parts, the central ethene bridge and aryl at side chain, which undergo a ring-closure reaction between a colourless open form and a coloured closed form. Although "aryl" may include other aromatic structures, most of the DAEs are based on heterocycles. The typical chemical structure and reaction scheme of DAEs with heterocycle group are shown in Fig.5.10(a).

Upon irradiation with ultraviolet (UV) light, azobenzene **1** and spiropyran **2** convert from left-side isomers to right-side ones. The photogenerated right-side isomers are thermally unstable, and the colours disappear in the dark at room temperature. These traditional molecules are classified into T-type (thermally reversible) photochromic molecules. In contrast, such as fulgide **3** and diarylethene **4,** the photogenerated right-side isomers are thermally stable and hardly return to the left-side isomers in the dark at room temperature, these molecules are classified into P-type (thermally irreversible, but photochemically reversible) photochromic ones. The thermal irreversibility is an essential and indispensable property for the use of photochromic molecules in optical memories, switches, and molecular machines. For such applications, photochromic molecules should fulfil the following requirements: ①thermal stability of both isomers, ②fatigue-resistant property (high number of colouration/decolouration cycles), ③high sensitivity, ④rapid response, and ⑤reactivity in solid state.

The best performance of the DAEs is summarized as follows. ①Both isomers are thermally stable: well-designed derivatives have a half-life time at room temperature longer than 400000 years. ②Colouration/decolouration cycles can be repeated more than 10000 times. ③The quantum yield of colouration is close to 1 (100%). ④Both colouration and decolouration reactions take place in a picosecond time region. ⑤Many DAEs undergo photochromic reactions in the single crystalline phase. Due to above advantages, DAEs are considered as very serious candidates for optoelectronic applications (memories, switches, etc.). Thus, this unit will focus on DAEs.

(5) Development of DAEs

The DAEs are derivatives of stilbene. Stilbene is well-known to undergo a photocyclization reaction to produce dihydrophenanthrene. The dihydrophenanthrene returns to stilbene in the dark in a deaerated solution. In the presence of air, however, the dihydrophenanthrene irreversibly converts to phenanthrene by the hydrogen-elimination reaction with oxygen [Fig.5.10(b)].

When the 2- and 6-positions of the phenyl rings in stilbene are substituted with methyl groups, the elimination reaction is suppressed and the compound undergoes a reversible photocyclization reaction, that is, a photochromic reaction, even in the presence of oxygen. Although the introduction of methyl substituents at the 2 (or 6) and 2' (or 6') positions of the phenyl rings of compound **5** prevents the irreversible oxidation reaction, the photogenerated dihydro form (compound **5b**) is unstable and reverts quickly to the open-ring form [Fig.5.10(c)]. Such a thermally unstable photochromic system (T-type photochromic molecule) is not useful for optical memories and switches.

When phenyl rings of stilbene are replaced with five-membered heterocyclic rings with low aromatic stabilization energy, such as thiophene or furan rings, both open- and closed-ring isomers become thermally stable and colouration/decolouration cycles can be repeated many times. For example, as shown in Fig.5.10(d), the dicyano and maleic anhydride groups are selected to shift the absorption maxima of the dihydro-type isomers to longer wavelengths. The photogenerated dihydro-type isomers, **6b** and **7b**, never returned to the initial open-ring isomers in the dark for more than 3 months, even at 80℃, but readily regenerated the open-ring isomers by irradiation with visible light (λ > 450nm). These are the first examples of thermally irreversible photochromic diarylethenes (P-type photochromic molecule). Since then, various types of diarylethenes having thiophene, furan, indole, selenophene, and thiazole aryl groups have been prepared.

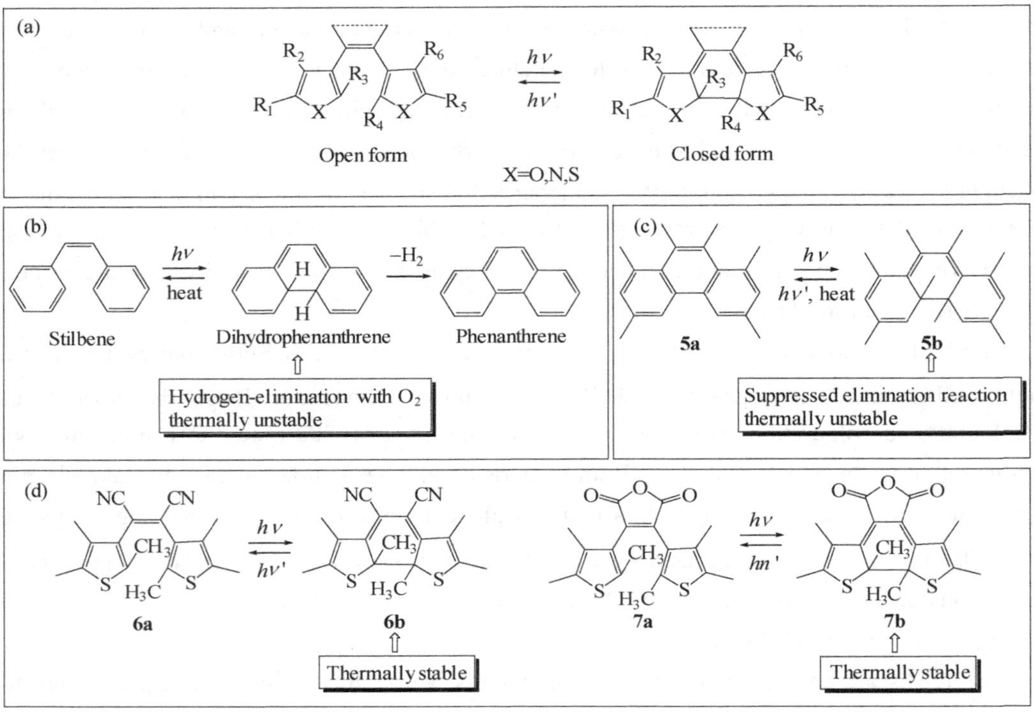

Fig.5.10 (a) Typical reaction scheme of the photochromism of diarylethene when the aryl group is a heterocycle; (b) The reaction stilbene undergo to produce dihydrophenanthrene; (c) The reaction compound 5a (open form) undergo to produce 5b (closed form); (d) First examples of thermally irreversible photochromic diarylethenes

3. DAEs Based on Benzobisthiadiazole as an Ethene Bridge

The central ethene bridges, which is necessary for the versatility of the DAE architectures, have been mostly limited to a cyclopentene unit or its strong electron-withdrawing analogues, such as perfluorocyclopentene [compound **8** in Fig.5.11(a)], maleic anhydride [compound **7** in Fig.5.10(d)], or maleic imide.

A photochromic system [BTTE in Fig.5.11(a)] based on the benzo[1,2-*c*:3,4-*c*′]bis[1,2,5] thiadiazole (abbreviated as benzobisthiadiazole) chromophore as a novel six-membered ring containing a central bridging ethene unit is reported, which is expected to have low aromaticity, and yield a thermally stable closed form (*c*-BTTE). This compound possesses several merits: ①excellent photochromic performance in both solution [Fig.5.11(b)] and single crystals; ②an extremely low aromaticity of the central ethene bridging unit of benzobisthiadiazole as a result of the high polarity and electron-withdrawing tendency of the benzobisthiadiazole unit; ③good thermal stability in a variety of solvents (ranging from nonpolar cyclohexane to polar ethanol) and ④a convenient modulation of the fluorescence of BTTE with photochromism and solvatochromism [Fig.5.11(c)].

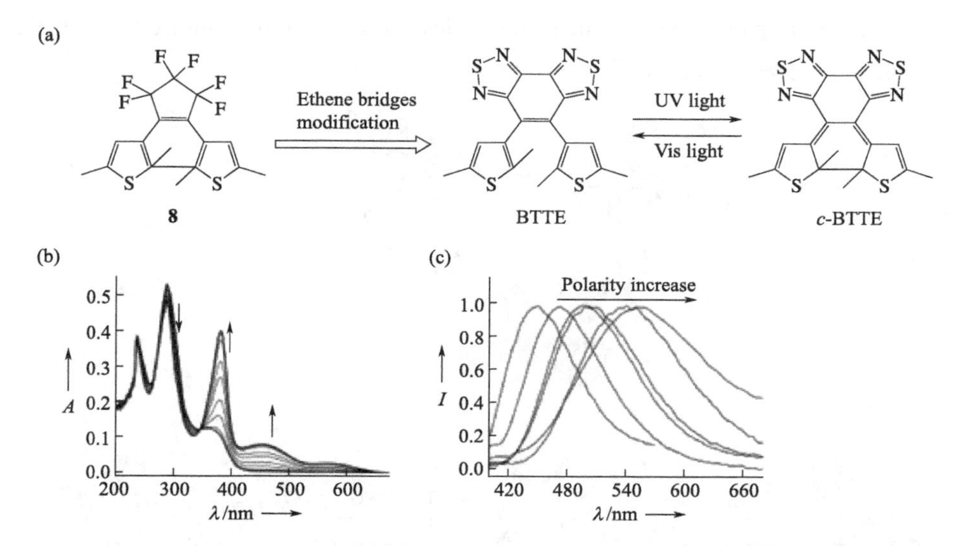

Fig.5.11 (a) Chemical structures of compound 8 and BTTE; (b) Changes in the absorption spectrum of BTTE upon irradiation at 365nm in THF (2.09×10^{-5}mol · L^{-1}) over 360s; (c) Normalized fluorescence spectra of BTTE in various solvents (from left to right: cyclohexane, chloroform, THF, acetonitrile, DMF, and ethanol)

Generally, diarylethenes bestow two conformers with two aryl rings in mirror symmetry (parallel conformation, *p*-) and C_2 symmetry (anti-parallel conformation, *ap*-). As the conrotatory reaction can only result from the anti-parallel conformation, the photocyclization quantum yield ($\Phi_{o\text{-}c}$) in conventional DAEs is usually limited to 50%. With critical concerns on the photo-switching efficiency and thermal irreversibility, it is pivotal to tailor the equilibrium between *p*- and *ap*-conformations for an increase in the photocyclization quantum yield.

Based on BTTE, a benzene ring is conjugated to the thiophene unit as a larger bulky terminal benzothiophene for further development of the thermally irreversible DAE system (Fig.5.12). The steric interaction between the bulky terminal benzothiophene and the benzobis thiadiazole bridge is so large that the rotation of benzothiophene is extremely hindered. The common interconversion between the *p*- and *ap*-conformers is not observed, and **9a** and **9p** as *ap*- and *p*-conformers are successfully separated with silica gel chromatography and recrystallization. As expected, $\Phi_{o\text{-}c}$ of **9a** is determined to be 72.9% in THF, which is much higher than that of BTTE (26.4 % in THF). The presence of ICT in **9a** is deleterious to the photocyclization owing to the existence of the competing deactivating channel in excited states. To further improve $\Phi_{o\text{-}c}$, the ICT effect is blocked by introducing strong electron-withdrawing nitro groups on the benzothiophenes, and $\Phi_{o\text{-}c}$ is as high as 90.6% with efficiently blocked ICT (Fig.5.12).

The work provides deep insight into how hindered conformation affects the photochromism of

DAEs, opening up a breakthrough to high $\Phi_{o\text{-}c}$ through both separating completely pure *ap*-conformers and suppressing ICT. DAEs based on benzobisthiadiazole as an ethene bridge enriched the family of DAEs. What's more, the outstanding performance of the diarylethenes offers great potential for advancing future optics and optoelectronic technologies and also certain promise for applications to biological science and technologies.

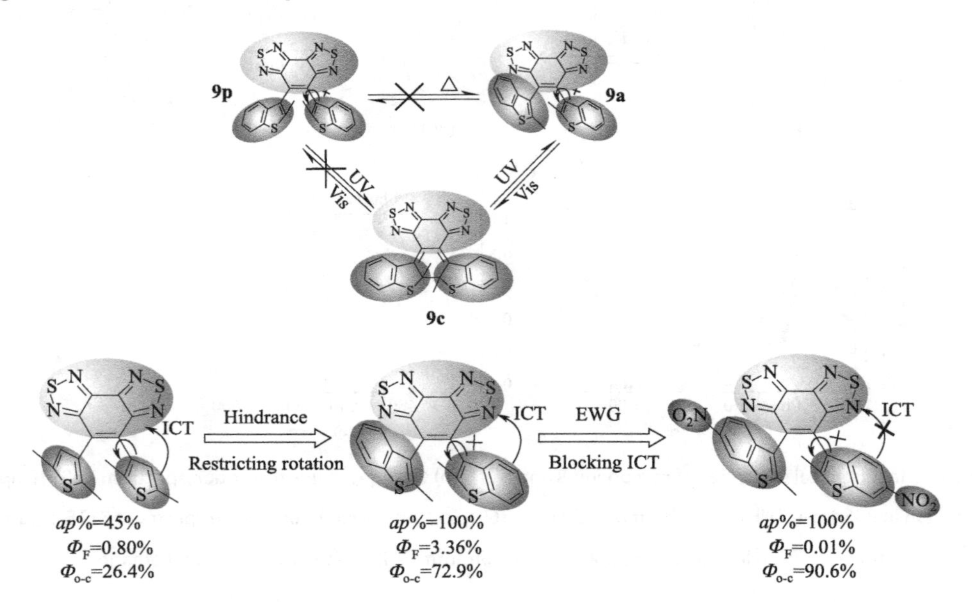

Fig.5.12 The conversion relationship of 1a, 1p, and 1c with a hindered aryl group, efficient modulation of $\Phi_{o\text{-}c}$ through separating completely pure *ap*-conformer and suppressing ICT

Words and Expressions

photochromism *n.* 光致变色
photochromic *adj.* 光致变色的
anti-counterfeit *v.* 防伪
super-resolution imaging 超分辨成像
diarylethene *n.* 二芳基乙烯
deaerate *v.* 除气
quantum yield 量子产率
fatigue resistance 抗疲劳性
photostationary state 光稳态
photocyclization *n.* 光环化反应
elimination *n.* 消除；除去
irreversible *adj.* 不可逆的
aromaticity *n.* 芳香性；芳香族化合物的结构
solvatochromism *n.* 溶致变色
parallel *adj.* 平行的
anti-parallel *adj.* 反平行的
conrotatory *adj.* 顺旋的
steric *adj.* 空间的；立体的

Notes

➢ When the 2- and 6-positions of the phenyl rings in stilbene are substituted with methyl groups, the elimination reaction is suppressed and the compound undergoes a reversible photocyclization reaction, that is, a photochromic reaction, even in the presence of oxygen.

参考译文：当对称二苯基乙烯中苯环的 2 位和 6 位被甲基取代时，消除反应被抑制。即使在氧气存在下，化合物仍然可以产生一个可逆的光环化反应，即光致变色反应。

➢ The steric interaction between the bulky terminal benzothiophene and the benzobis thiadiazole bridge is so large that the rotation of benzothiophene is extremely hindered.

参考译文：末端大体积苯并噻吩基团与苯并二噻二唑烯桥的空间相互作用很大，以至于苯并噻吩基团的旋转受到极大阻碍。

Exercises

1. Discuss the following questions

(1) What are T-type photochromic molecules?

(2) Many small organic molecules undergo irreversible photobleaching under high-intensity light (e.g., green solution of ICG become colourless after irradiation of light), is this phenomenon classified into photochromism? Why?

(3) Why most of the DAEs are based on heterocycles?

(4) What makes compound **9a** have such a high $\Phi_{o\text{-}c}$ of 90.6%?

2. Translate the following into Chinese

(1) Photochromism is a reversible transformation of a chemical species induced in one or both directions by absorption of electromagnetic radiation between two forms, A and B, having different absorption spectra.

(2) UV light irradiation of the coulorless closed form leads to the carbon-oxygen bond cleavage (open ring reaction) of the pyran ring.

(3) Conjugation is broken at the spiro carbon atom, and the molecule is colourless and absorbs only in the UV region.

(4) Chemical structure of diarylethenes (DAEs) consists of two parts, the central ethene bridge and aryl at side chain, which undergo a ring-closure reaction between a colourless open form and a coloured closed form.

(5) The photogenerated dihydro-type isomers, **6b** and **7b**, never returned to the initial open-ring isomers in the dark for more than 3 months, even at 80℃, but readily regenerated the open-ring isomers by irradiation with visible light.

(6) Generally, diarylethenes bestow two conformers with two aryl rings in mirror symmetry (parallel conformation, *p*-) and C_2 symmetry (anti-parallel conformation, *ap*-).

3. Translate the following into English

(1) 与之相反，光致变色反应引起了分子自由体积的显著变化。

(2) 虽然“芳基”可能包含其他芳香结构，但大多数二芳基乙烯是基于杂环的。

(3) 在室温下，光照产生的右侧同分异构体是热稳定的，并且黑暗下几乎不回到左侧的同分异构体。

(4) 这两种异构体都是热稳定的：精心设计的衍生物在室温下的半衰期超过 40 万年。

(5) 通过硅胶色谱和重结晶，反平行和平行构象（**9a** 和 **9p**）被成功分离。

UNIT 3 Dye-sensitized Solar Cells

1. Introduction

Solar energy is regarded as one of the perfect energy resources owing to its huge reserves, inexhaustibility and pollution-free character. Directly converting solar light to electric energy based on photovoltaics is the optimal way for electrified modern society. Different technologies in photovoltaics, such as crystalline Si, semiconductor (e.g., GaAs)-based cells, thin-film (e.g., CdTe) solar cells, organic bulk heterojunction (BHJ) solar cells and dye-sensitized solar cells (DSSCs), coexist to compete for the future market. Among them, the dye-sensitized solar cells (DSSCs) have been considered as a promising generation of photovoltaic techniques due to their unique biomimetic operating principle with low cost and high efficiency.

2. Device Structure, Working Principle and Fabrication Process of Dye-sensitized Solar Cells (DSSCs)

(1) Device Structure

A general dye-sensitized solar cell is mainly composed of five parts: transparent conducting oxide (TCO) substrate, a mesoporous semiconductor (predominately TiO_2) film adsorbed with photosensitizer dyes, an electrolyte layer containing a redox couple, and a counter electrode, as shown in Fig.5.13. The core part of DSSCs is a porous semiconductor film composed of nano-TiO_2 particles. The diameter of the TiO_2 particles is about 10~30nm, the thickness of the film is about 10μm, and the porosity is 50%~60%. In the process of preparing the film, the nano-TiO_2 is sintered to form a closely connected network structure between particles, which not only provides a huge specific surface area, but also has good electronic conductivity between particles. Nano-semiconductor films are generally attached to a transparent conductive substrate, and the most commonly used substrate is FTO conductive glass. A layer of photosensitive dye with charge transfer function is chemically adsorbed on the nano-semiconductor film. Because the membrane has a porous structure and the pore size is much larger than the volume of the dye molecules, the photosensitive dye can

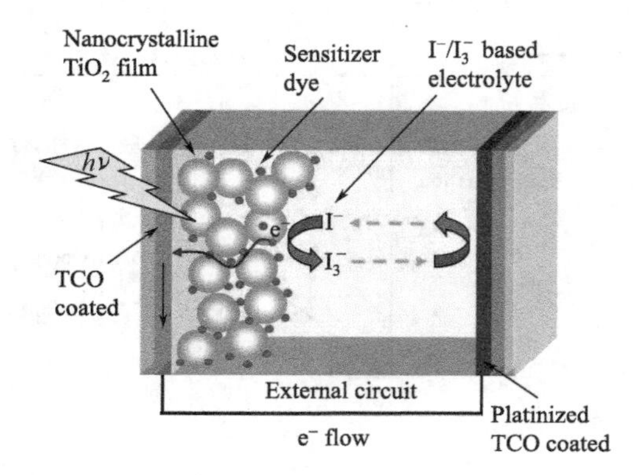

Fig.5.13 Schematic diagram of dye-sensitized solar cells (DSSCs)

not only adsorb on the membrane surface, but also enter the membrane through the pores and adsorb on its huge inner surface. The nano-semiconductor film attached to the conductive substrate forms a photoanode after adsorbing photosensitive dye. The photoanode and the counter electrode sandwich the electrolyte to form a sandwich structure. The thickness of the electrolyte is about 30~50μm. The more commonly used electrolyte is I^-/I_3^- redox couple and some additives with nitrile organic solution. The counter electrode is mostly platinum-plated conductive glass.

(2) Working Principle

As shown in Fig.5.14, a series of kinetic processes and chemical reactions of electron transfer occur inside DSSCs when they work, mainly including:

① Excitation of photosensitizer dye　The dye sensitizer adsorbed on the TiO_2 particles absorbs incident photons and transitions from the ground state to the excited state. Generally speaking, the excited state lifetime is 10^{-9}s, which is nanosecond level. The longer the lifetime of the excited state, the better the subsequent electron injection.

② Electron injection　The excited state dye injects electrons into the conduction band of the TiO_2 semiconductor. This process is very fast, usually at the picosecond level (10^{-12}s).

③ Dye regeneration　The oxidized dye with poor electrons obtains electrons from the reduced pair in the electrolyte and returns to an electrically neutral state. The reduced pair loses electrons and is oxidized. This process generally occurs in microseconds (10^{-6}s).

④ Charge collection　The electrons injected into the conduction band of TiO_2 pass through the network structure and are collected on the FTO of the photoanode. This process is about 100 microseconds. In this process, the electrons may recombine with the dye in the oxidation state or the reduced pair in the electrolyte. Finally, the successfully collected electrons enter the external circuit and flow to the counter electrode after passing through the load.

⑤ Electrolyte regeneration　The oxidized reducing couple diffuses to the counter electrode, and the electrons are obtained by catalytic reduction on the counter electrode, thus completing the entire circuit.

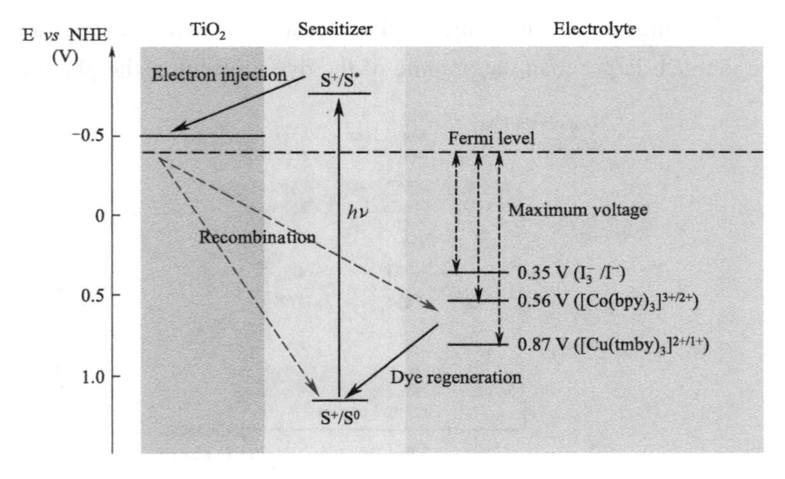

Fig.5.14　Operating principles and energy level diagram of dye-sensitized solar cells

(3) Fabrication Process

The production of DSSCs can be divided into three processes: photoanode preparation, counter electrode preparation and battery assembly (Fig.5.15).

Preparation of photoanode: First, wash the FTO with detergent, deionized water, acetone and ethanol in sequence. Use a semi-automatic screen printing or manual screen printing machine to coat a uniform mesoporous TiO_2 slurry on the surface of the FTO. The film thickness can be controlled by the mesh size of the screen and the pressure of the coating film, and the desired film thickness can also be obtained by printing multiple times. The prepared film is immersed in the dye solution for sensitization. Before this, the dye solvent needs to be optimized. And the more commonly used solvent system is a binary solvent mixed with a good solvent and a poor solvent.

Preparation of counter electrode: Firstly, small holes are made on the conductive FTO, and then the corresponding preparation method of the catalytic material is selected according to the different redox couples. The general preparation methods include spin coating, electrodeposition, sputtering and so on.

Battery assembly: The photoanode and counter electrode are pressurized and packaged with hot melt adhesive at a temperature of about 120℃. Finally, the prepared electrolyte solution is injected through the small hole on the counter electrode, and the small hole is sealed with a glass sheet and hot melt glue.

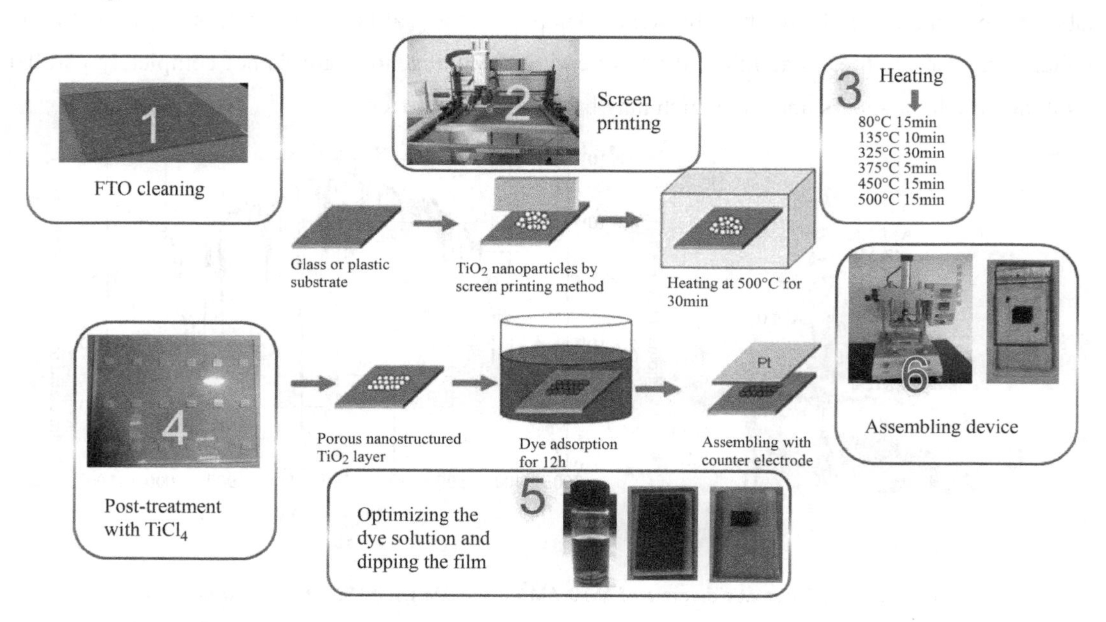

Fig.5.15 Fabrication technique of dye-sensitized solar cells

3. Performance Evaluation of DSSCs

The performance evaluation of solar cells mainly includes two aspects: photoelectric conversion efficiency and device stability. Here, we will introduce the current research methods for evaluating battery efficiency and stability, and analyze the energy loss in DSSCs.

(1) Solar Energy and Solar Spectrum

To understand the photoelectric conversion efficiency of solar cells, it is first necessary to understand the nature of its energy source—the sun. The spectrum radiated by the sun is a composite spectrum with a wide range of wavelengths (including ultraviolet light at 280nm to infrared light at 4000nm), and the photon flux density at each wavelength is also different. When sunlight reaches the surface of the earth, the atmosphere has a certain weakening effect on the intensity of sunlight. The intensity of solar radiation received by different locations on the earth and at different times is different, so studying the photoelectric conversion efficiency of solar cells requires a uniformly recognized standard light source.

Generally, air mass (AM) is used to define the degree of influence of the atmosphere on the solar energy received on the earth's surface. When there is no atmosphere, it is called AM 0. When the sunlight penetrates the atmosphere and illuminates the ground vertically, it is AM 1. When the angle between the sun's rays and the ground is θ, the air mass is AM = $1/\cos\theta$. In particular, when $\theta = 48.2°$, the air mass is AM 1.5 [Fig.5.16(a)]. Sunlight with this angle of incidence is common in most parts of the world. Therefore, in solar cell research, AM 1.5 light source is generally used as the standard light source [Fig.5.16(b)] for solar cell testing, and the corresponding light intensity is $1000W \cdot m^{-2}$ ($100mW \cdot cm^{-2}$). Xenon lamps are commonly used as simulated solar light sources in laboratories, but the light emitted by xenon lamps is easily affected by factors such as driving voltage and service life. In addition, the light emitted by the xenon lamp is not completely parallel light, and the light spot is uneven, which may bring errors to the test.

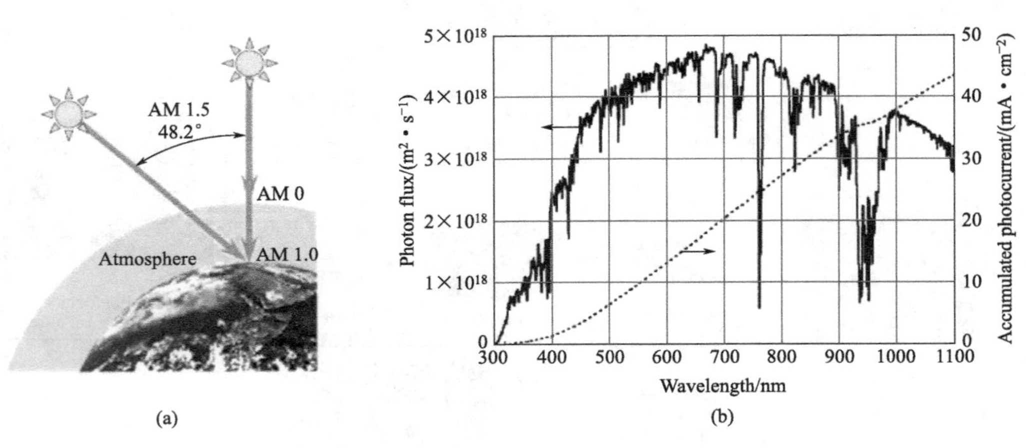

Figure 5.16 (a) Schematic diagram of AM 0, AM 1.0 and AM 1.5; (b) Spectrum distribution of AM 1.5 solar irradiation (theoretical J_{sc} vs. wavelength)

(2) Current-voltage Characteristics (*I-V*) and Photoelectric Conversion Efficiency η of Solar Cells

The current-voltage output characteristics of solar cells are shown in the curve in Fig.5.17. The abscissa in the figure is the battery output photovoltage, and the ordinate is the output photocurrent.

In the test, the photovoltage gradually increases from zero until the photocurrent drops to zero. The entire curve is composed of a series of photocurrent-photovoltage data points. Among them, the data at the following three points are of more concern:

① When the photovoltage $V = 0$, it is a short-circuit state, and the photocurrent is called the short-circuit current I_{sc};

② When the photocurrent $I = 0$, it is an open-circuit state, and the photovoltage at this time is called the open-circuit voltage V_{oc};

③ The product of photovoltage and photocurrent, that is, when the output power P is maximum, the corresponding photocurrent is called I_{max}, and the photovoltage is called V_{max} ($P_{max} = I_{max} \times V_{max}$).

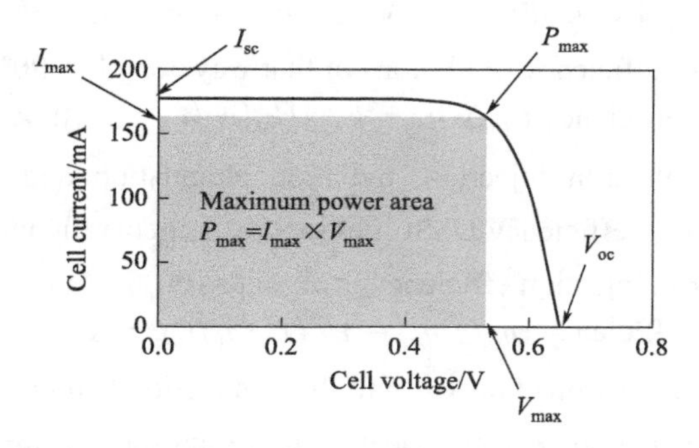

Fig.5.17 Schematic $I-V$ curve of DSSCs

The ratio of the maximum battery output power (P_{max}) to the theoretical maximum power ($I_{sc} \times V_{oc}$) is called the fill factor (FF) (formula 5.1):

$$FF = \frac{I_{max} \times V_{max}}{I_{sc} \times V_{oc}} \tag{5.1}$$

The fill factor is an important parameter to measure the quality of solar cell devices. The main factors affecting the battery FF are the internal series resistance of the battery: including charge transfer impedance in the semiconductor and electrolyte, transfer impedance at the counter electrode/electrolyte interface, and surface impedance of the conductive substrate. Reducing these impedances can effectively increase the FF value.

The efficiency η of the solar cells can be calculated from the data of the above three points and the incident light power P_{in}

$$\eta = \frac{P_{max}}{P_{in}} = \frac{I_{sc} \times V_{oc} \times FF}{P_{in}} \tag{5.2}$$

It can be found from formula 5.2 that the photoelectric conversion efficiency of solar cells is mainly determined by short-circuit photocurrent I_{sc} (after normalizing the battery area, it is called the short-circuit photocurrent density J_{sc}, and the unit is $mA \cdot cm^{-2}$), open-circuit photovoltage V_{oc} and fill factor FF.

(3) Quantum Efficiency, Incident Photon to Current Conversion Efficiency (IPCE) and Short-circuit Photocurrent Density (J_{sc}) of Solar Cells

From a quantum point of view, a solar cell is a tool that can convert a flow of photons into electrons (or current). The quantum efficiency of solar cells is the ratio of the number of electrons contributed by the solar cells to the external circuit to the number of incident photons. The conversion process of photons to electrons in DSSCs mainly involves three processes: photon absorption, electron injection and electron collection, and each step corresponds to its quantum efficiency.

① The photon absorption is expressed by the light harvesting efficiency LHE, LHE = $10^{-\alpha}$~1, and α is the absorbance of the photon of the wavelength λ by the battery. The photons in DSSCs are mainly absorbed by dyes, while most of the dyes are distributed in unevenly band absorption, so the LHE at different wavelengths is different. When the absorbance of the battery at a certain wavelength is 1, the capture efficiency of photons of that wavelength is 90%.

② Electron injection efficiency (η_{inj}): $\eta_{inj} = k_{inj} / (k_{inj} + k_{rad} + k_{nrad})$, k_{inj}, k_{rad} and k_{nrad} correspond to the excited state dye electron injection, radiation attenuation and non-radiation attenuation constants respectively. In high-efficiency DSSC devices, k_{inj} is generally more than 100 of the sum of k_{rad} and k_{nrad}, and the electron injection efficiency is close to 100%.

③ Charge collection efficiency (η_{col}): $\eta_{col} = 1 / (1 + \tau_t / \tau_n)$, τ_t is the electron transit time, and τ_n is the electron lifetime. The electrons injected into the conduction band of TiO_2 will recombine with the oxidized dye or the reduced pair in the electrolyte during the transfer to the photoanode FTO. When the electron lifetime τ_n is much longer than the electron transfer time τ_t, the electrons are not easy to recombine, and the charge collection efficiency is higher. τ_t and τ_n can be obtained by electrochemical AC impedance or transient photovoltage decay test.

Considering that η_{inj} and η_{col} are both closely related to the electron concentration in the conduction band of TiO_2, that is, it is closely related to the photovoltage. Generally, the quantum efficiency of the battery in the short-circuit state is mainly studied, in this case it is called the incident photon to current conversion efficiency (IPCE) (formula 5.3)

$$\text{IPCE} = \text{LHE} \times \eta_{inj} \times \eta_{col} \tag{5.3}$$

In the actual test, a monochromator is generally used to decompose the simulated sunlight into monochromatic light of continuous wavelength and irradiate the surface of the battery. Recording the short-circuit current value at each wavelength, and using formula 5.4 to get the IPCE value at different wavelengths. The IPCE values at different wavelengths are connected to form the IPCE curve, so the integral of the IPCE curve is the short-circuit current density (J_{sc}) of the battery

$$\text{IPCE} = \frac{J_{sc}(\lambda)}{e\Phi(\lambda)} = 1240 \frac{J_{sc}(\lambda)[\text{A} \cdot \text{cm}^{-2}]}{\lambda[\text{nm}]P_{in}[\text{W} \cdot \text{cm}^{-2}]} \tag{5.4}$$

When designing DSSC devices, in order to obtain a larger J_{sc} value, it is necessary to start with the width and height of the IPCE at the same time. On the one hand, a sensitizer with a broad spectrum and high molar extinction coefficient response is designed to obtain better light capture efficiency (LHE) and ideal IPCE width. On the other hand, it is necessary to ensure that the LUMO and HOMO energy levels of the dye have a sufficient energy level difference with the Nernst potential of the TiO_2 conduction band and the electrolyte, as the driving force for electron injection and dye regeneration, and ensuring sufficient η_{inj} and η_{reg} to obtain the ideal IPCE platform. In addition, the configuration of the dye molecule will affect η_{col} to a certain extent, which in turn affects the height of the IPCE platform. In general, the design of DSSC devices with high short-circuit current needs to consider these factors comprehensively, and reasonably adjust the energy levels of sensitizers, metal semiconductors, and redox media.

(4) Open-circuit Voltage (V_{oc}) of DSSCs

The principle of photovoltage generation in DSSCs is that the electrons injected into TiO_2 by the sensitizer accumulate in the conduction band and promote the rise of its Fermi level. The open-circuit voltage value (V_{oc}) is determined by the energy level difference between the Fermi level of TiO_2 and the redox potential of the electrolyte, which can be expressed as formula 5.5

$$V_{oc} = \frac{E_{CB}}{q} + \frac{kT}{q}\ln\left(\frac{n}{N_{CB}}\right) - \frac{E_{redox}}{q} \tag{5.5}$$

E_{CB} is the energy level of the conduction band of the metal oxide, k is the Boltzmann constant, n is the number of photoelectrons in the metal oxide, and N_{CB} is the number of all conduction band excited states in the metal oxide. Therefore, increasing V_{oc} can be considered from three aspects: increasing the conduction band energy level E_{CB} and the number of photoelectrons n, or reducing E_{redox}.

Many researchers put their emphasis on the sensitizer dyes because they play a key role in DSSCs: ①their absorption properties (including the wavelength range and extinction coefficients) determine the light-harvesting of DSSCs; ②their levels of molecular frontier orbitals influence thermodynamics in electron injection and dye regeneration; ③their configuration and aggregation state on semiconductor surfaces affect the competition between electron injection and recombination; and ④their photothermal stability determines the device lifetime to a great extent. The correlation between molecular sensitizer structure and device efficiency is an important issue in DSSC researches.

4. Research Progress of DSSCs

Dye-sensitized solar cells based on nanocrystalline thin films have been extensively studied by many research groups around the world since they were reported in 1991. It has made great progresses mainly in three aspects: efficiency, stability and understanding of the mechanism. Every progress and breakthrough is accompanied by the innovation of materials and the re-examination of the mechanism. Dye studied photo-chemists and synthetic-chemists have improved the structure of dyes in terms of energy level matching, spectral width, molecular configuration and aggregation state for improving battery efficiency and stability.

Photosensitive dyes have always been a hot topic in DSSC researches. The structural modification of dyes can effectively broaden the spectral response of DSSCs and improve the photoelectric efficiency and stability. Presently, organic sensitizers for DSSCs fall into two broad categories: metal-polypyridyl complexes and pure metal-free organic dyes. DSSCs based on ruthenium (Ru)-polypyridyl dyes usually show high efficiency [Fig.5.18(a)]. However, with a focus on the cost and limited Ru resource, a metal-free organic sensitizer becomes very promising. Undoubtedly, molecular engineering is the most effective method to improve the performance from the viewpoint of sensitizers.

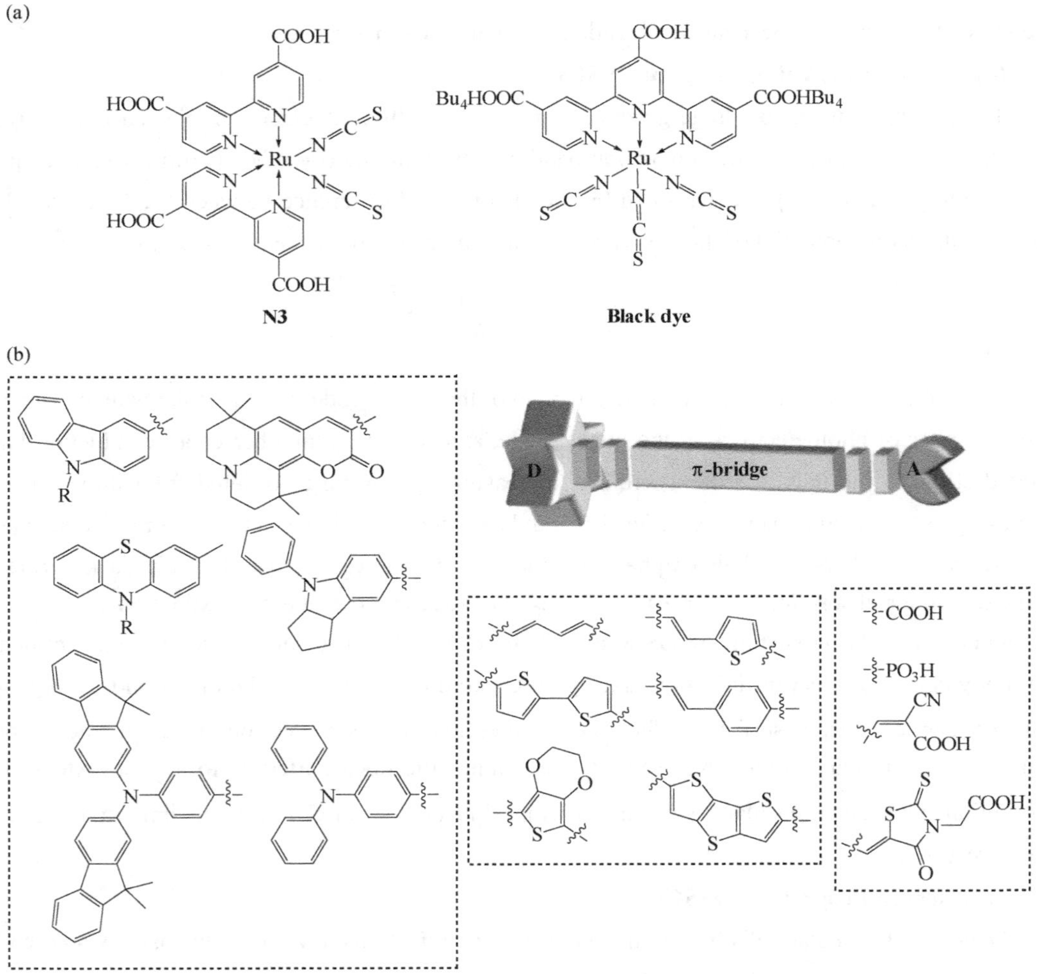

(a)

N3

Black dye

(b)

Fig.5.18 (a) Chemical structures of classical Ru dye N3 and black dye; (b) Constructing pure organic sensitizers by using D-π-A model and some common structures of electron donor, acceptor and conjugating units

From 2001 to 2007, most of these pure organic sensitizers have been constructed with a typical electron donor-π bridge-electron acceptor (D-π-A) configuration [Fig.5.18(b)] for improving J_{sc} and V_{oc} through regulating HOMO-LUMO energy levels of dye molecules or the adsorption, arrangement and aggregation state of dyes on the surface of TiO_2. Adjusting the structure of the donor and acceptor and extending the length of the molecular conjugate chain are the means of regulating the spectrum and energy level of D-π-A dye. Among them, extending the conjugation is the main means to broaden the spectrum of the dye. However, the longer conjugate chain will easily lead to the aggregation of the dye, and it is not conducive to the photothermal stability of cells. Meanwhile, dye absorption spectrum broadening and electron transferring are contradictory in energy levels. Broadening the dye absorption spectrum will reduce the LUMO energy or raise the HOMO energy level, which will be accompanied by a reduction in the driving force for electron transfer, resulting in a significant drop in battery performance. Therefore, there is a delicate balance between dye absorption spectrum broadening and electron transferring, and this is the core of the dye design.

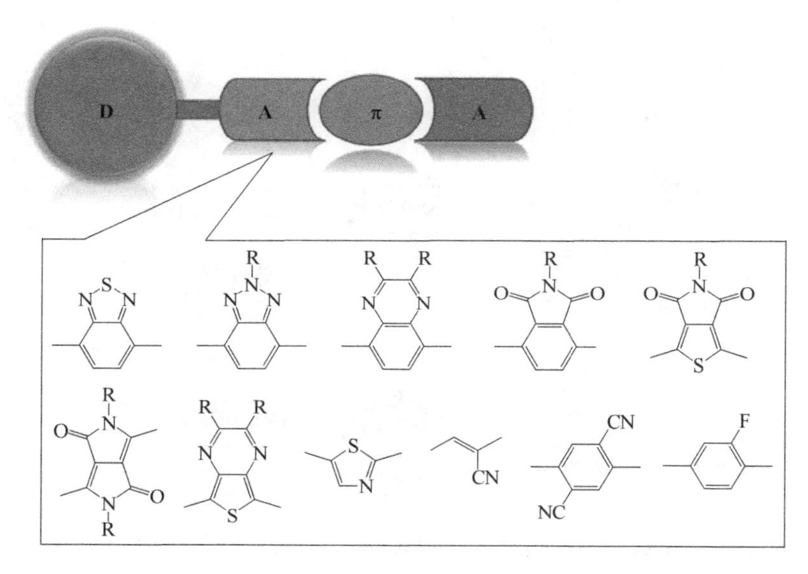

Fig.5.19 Configuration of organic D-A-π-A sensitizers as well as building blocks for additional electron-withdrawing acceptors

To overcome the intrinsic drawbacks of traditional D-π-A, recently, a novel strategy, called D–A–π–A, has been proposed (Fig.5.19) for designing novel organic sensitizers, in which several kinds of electron-withdrawing units (such as benzothiadiazole, benzotriazole, quinoxaline, phthalimide and diketopyrrolopyrrole) are incorporated into the π bridge to tailor molecular structures and optimize energy levels. It has been systematically demonstrated that the incorporated electron-withdrawing additional acceptor can be treated as an "electron trap", showing several distinguished merits such as: ①essentially facilitating the electron transfer from the donor to the acceptor/anchor; ②conveniently tailoring the solar cells performance with a facile structural modification on the additional acceptor; ③improving V_{oc} with the nitrogen-containing heterocyclic group; ④conveniently tuning the molecular energy gap, and modulating the response of the light-harvesting range with the new resulting absorption band; and ⑤most importantly, being capable of greatly improving the sensitizer photo-stability.

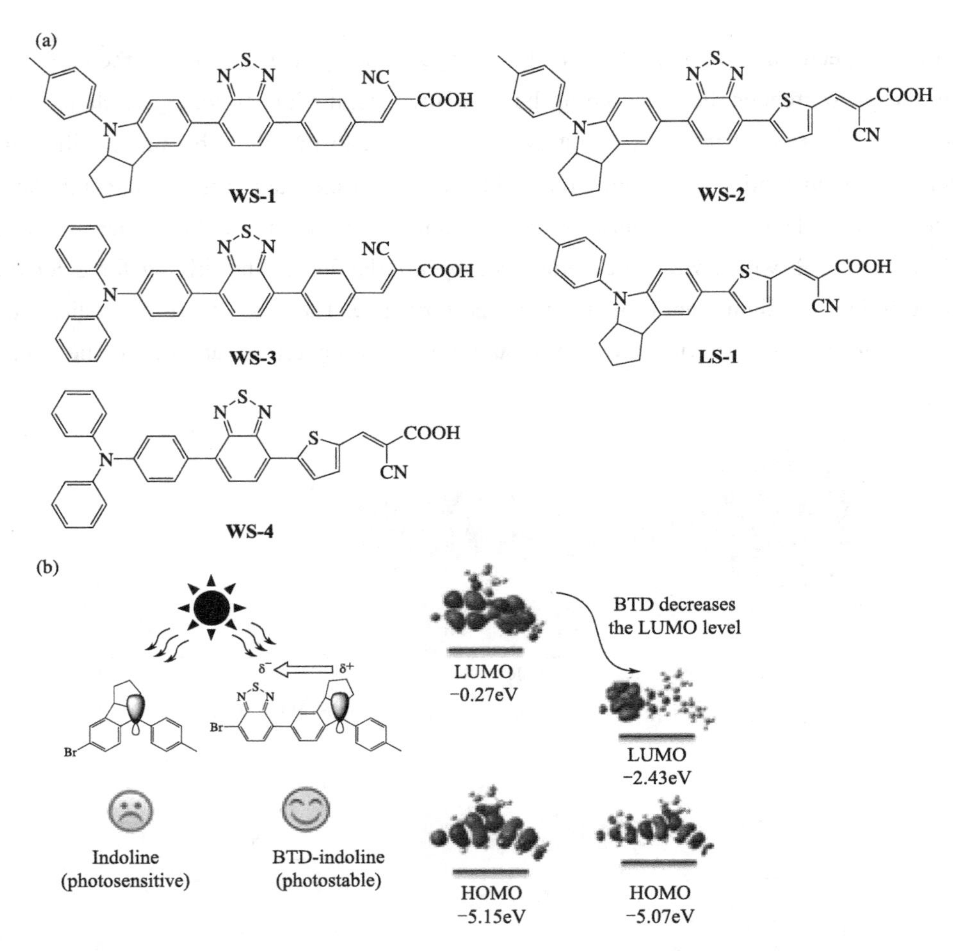

Fig.5.20 (a) Chemical structures of BTD-based D-A-π-A featured organic sensitizers WS-1 ~ WS-4 and reference dye LS-1; (b) Mechanism of the BTD unit for increasing the photo-stability of indoline derivatives

As an important building block, BTD [Fig.5.20(a)] is a strong electron-withdrawing unit with a five-membered heterocyclic ring (C═N—S—N═C), exhibiting the electron-deficient character. It has been demonstrated the contribution of the BTD-based D-A-π-A configuration to the energy level [Fig.5.20(b)] optimization and light harvesting enhancement (red-shift of the CT band, generation of additional absorption bands, decrease of the hypsochromic effect on TiO_2 film, and improvement of the stability of indoline-based organic sensitizers), especially with respect to the common D-π-A configuration.

Words and Expressions

solar energy 太阳能

inexhaustibility *n.* 取之不尽

pollution-free *adj.* 无污染的

photovoltaics *n.* 光伏

dye-sensitized solar cell 染料敏化太阳能电池

biomimetic *adj.* 仿生的
device *n.* 设备; 器件
mesoporous *n.* 介孔
semiconductor *n.* 半导体
photosensitizer *n.* 光敏剂
conduction band 导带
external circuit 外电路
electrolyte *n.* 电解液
electrode *n.* 电极

Notes

➢ Solar energy is regarded as one of the perfect energy resources owing to its huge reserves, inexhaustibility and pollution-free character.

参考译文：太阳能因其储量巨大、取之不尽、无污染等特点被认为是最理想的能源之一。

➢ In brief, such solar cell devices are composed of a transparent conducting oxide (TCO) substrate, a mesoporous semiconductor (predominately TiO_2) film adsorbed with photosensitizer dyes, an electrolyte layer containing a redox couple, and a counter electrode.

参考译文：简而言之，这种太阳能电池器件由透明导电氧化物(TCO)衬底、被光敏剂染料吸附的介孔半导体(主要为 TiO_2)薄膜、含有氧化还原偶联的电解质层和一对电极组成。

➢ It has been demonstrated the contribution of the BTD-based D-A-π-A configuration to the energy level optimization and light harvesting enhancement (red-shift of the CT band, generation of additional absorption bands, decrease of the hypsochromic effect on TiO_2 film, and improvement of the stability of indoline-based organic sensitizers), especially with respect to the common D-π-A configuration.

参考译文：与常见的 D-π-A 结构相比较，基于 BTD 的 D-A-π-A 构型对能量级优化和光捕获增强具有贡献(CT 带红移，产生额外的吸收带，降低 TiO_2 薄膜的浅色效应，改善吲哚基有机增敏剂的稳定性)。

Exercises

1. Discuss the following questions

(1) What is the working principle of DSSCs?

(2) What is the advantage and disadvantage of ruthenium (Ru)-polypyridyl dyes?

(3) What is the advantage of BTD-based D-A-π-A configuration compared to D-π-A?

(4) What is the main structure of dye-sensitized solar cell device?

2. Translate the following into Chinese

(1) A general dye-sensitized solar cell is mainly composed of five parts: transparent conducting oxide (TCO) substrate, a mesoporous semiconductor (predominately TiO_2) film adsorbed with photosensitizer dyes, an electrolyte layer containing a redox couple, and a counter electrode.

(2) From a quantum point of view, a solar cell is a tool that can convert a flow of photons into electrons (or current).

(3) The structural modification of dyes can effectively broaden the spectral response of DSSCs and improve the photoelectric efficiency and stability.

(4) Undoubtedly, molecular engineering is the most effective method to improve the performance from the viewpoint of sensitizers.

3. Translate the following into English

(1) 光敏剂染料的激发:吸附在 TiO_2 粒子上的染料敏化剂吸收光子,从基态跃迁到激发态。

(2) 电子注入：激发态染料向 TiO_2 半导体的导带注入电子。

(3) 染料再生：缺电子的氧化染料从电解质的还原对中获得电子，并返回到电中性状态。

(4) 太阳能电池的量子效率是太阳能电池对外部电路贡献的电子数与入射光子数的比值。

UNIT 4 Molecular Logic-based Computation Systems

1. Introduction

Molecular logic-based computation is not about computational chemistry, and computational chemistry is the subject that depends on computer programs based on quantum theory running on semiconductor-based hardware to provide information about atoms, molecules, materials and reactions. However, molecular logic-based computation is about molecules and chemical systems which possess innate ability to compute, at least in a rudimentary way, like machines based on semiconductor transistors (or magnetic relays or mechanical abacuses in the past) or like people. As the semiconductor-based revolution gradually deals with ever-smaller features, molecules become ever more attractive as information processors.

Indeed, small molecules easily go to small vital spaces where semiconductor devices fear to tread. It therefore becomes a responsibility for chemists to explore the information-processing capabilities of molecules. This should not be difficult, because chemists have been exposed to molecular information processing since high school. Many chemistry experiments involve the exposure of a compound to a reagent and/or heat. Similar operations, perhaps with less quantitation, occur in kitchens around the world several times each day. Physical organic chemists consider the key compound to be the substrate. The progress of the reaction is seen by the change of some visible property to that of the product, such as the colour. This response of the compound to the stimulus can be appreciated in a different way (Fig.5.21). To borrow a computer scientist's language, an input is applied to a molecular device so that an output will result.

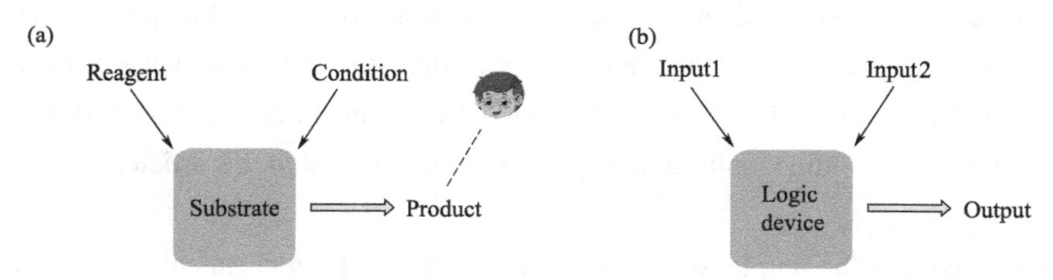

Fig.5.21 (a) Chemical and (b) computational aspects of stimulus-response situations

Electronic logic devices are characterized by easy application of inputs and equally easy observation of the outputs. We need to have similarly convenient operation of inputs/outputs to and from the molecular logic devices. Chemical inputs are easy to apply in a laboratory context. Optical responses (absorbance, emission intensity, emission anisotropy, ellipticity signals in circular dichroism spectra) serve as outputs since these are the most convenient for quick observation with inexpensive equipment or, sometimes, none at all. Of these, absorbance in the visible region is particularly convenient and visually striking. Emission intensity in the visible region is perhaps even more arresting, nearly as convenient and, furthermore, is detectable at the single molecule level. Of course, other spectroscopic signals arising from products of equilibria (nuclear magnetic resonance, Raman...) are also usable in this way. All of these circumvent problems of wiring between the molecular and macro worlds. Moleculebased materials also fit here even though they cannot result in single-molecule logic behaviour. Nevertheless, single-molecule logic is not essential for some imagined applications.

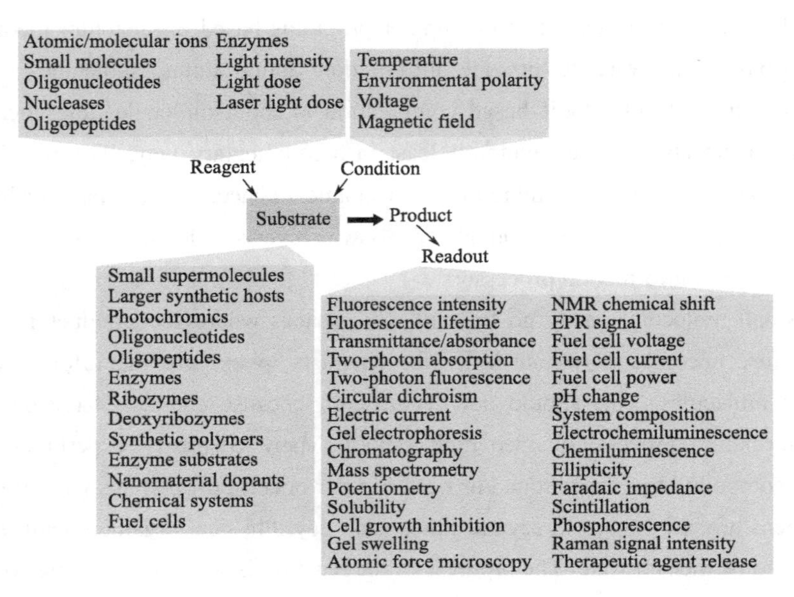

Fig.5.22 The range of inputs, outputs and molecular devices employed in logic operations

Optical inputs, e.g., light dose, which cause photochemical reactions, are also useful. “High” and “low” signal levels can be taken as binary 1 and 0 respectively (or the opposite) whatever the inputs or outputs under discussion. There are also molecular situations in which electric voltages are used as input and output signals. Many other chemical phenomena can be co-opted into molecular logic signals (Fig.5.22). Simple input-output systems are introduced in this article.

2. YES

YES logic requires the output to follow the input. Many of the chemical cases to be discussed below represent “off–on” switching of the output of some description, i.e., the application of a “high” input signal enhances the output signal.

Sensors are forms of YES or NOT logic gates. Protons may be the simplest chemical species, but they can be a convenient proving ground for many concepts. For instance, high levels of H^+ trigger strong fluorescence from compound **1** whereas there is almost no emission if H^+ concentrations are kept low. Photoinduced electron transfer (PET) occurs from the amine lone electron pair to the fluorophore in the latter situation. Arrival of high enough levels of H^+ blocks the lone electron pair of the amine and lets fluorescence reassert itself [Fig.5.23(a)].

A light-up fluorescent probe (BODIPY-β-gal) for sensing of β-galactosidase (β-gal) is developed. In this probe, monostyryl-substituted BODIPY (BODIPY-OH) is utilized as an near-infrared region (NIR) and wavelength-controllable chromophore, and a β-gal cleavable unit (β-galactopyranoside) as an enzyme-active trigger. There is a new absorption peak appeared at 620nm and a significantly enhanced fluorescence emission peak at 730nm in the presence of β-gal because of enhanced intramolecular charge transfer (ICT) process after cleavage [Fig.5.23(b)]. Absorption at 620nm or fluorescence emission at 730nm could serve as a "high" output signal of YES logic gates.

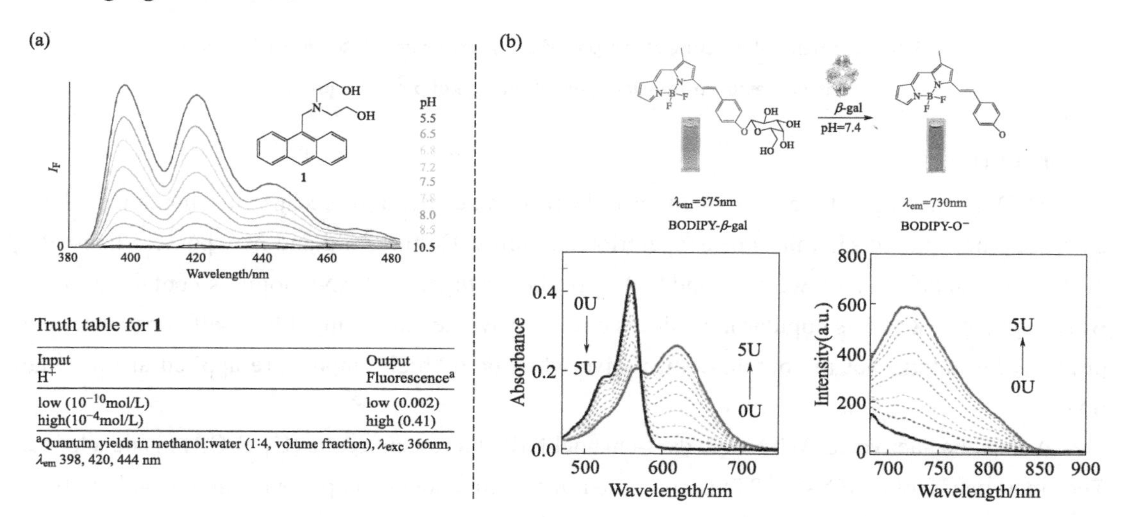

Fig.5.23 (a) Top: fluorescence intensity of 1 in methanol : water (1 : 4, volume fraction) as a function of pH, excitation wavelength 366nm; bottom: truth table for 1. (b) Top: ratiometric tracking of endogenous β-gal activity using an activatable near-infrared probe BODIPY-β-gal; bottom: absorption and fluorescence spectra of BODIPY-β-gal (10mmol · L^{-1}) incubation with β-gal (5U)

3. NOT

NOT logic naturally has the output behaving opposite to the input. In contrast to YES gates, the non-electronic NOT logic devices can be discerned in many sensors and reagents whose absorbance or fluorescence responds to the input in an "on-off" manner. In other words, the application of a "high" input signal causes the output signal to fall.

Convenient source of chemically driven NOT gates is the sensor literature, particularly the analytical fluorescence literature. This is to be expected because the quenching of fluorescence is a common phenomenon. Compound **2** is a "fluorophore-spacer-receptor" PET system. Compound **2** contains a pyridine receptor for H^+, which becomes more reducible when bound [Fig.5.24(a)]. Therefore, compound **2** launches fluorescence-quenching PET from the fluorophore to the pyridine receptor only when H^+ arrives. A selective Cu^{2+} probe **3** is developed, which can form well defined complex with Cu^{2+}. The fluorescence is almost completely quenched upon addition of 4eq. of Cu^{2+} due to the paramagnetic quenching effect [Fig.5.24(b)]. The low fluorescence signal output corresponds to the high concentration of Cu^{2+}.

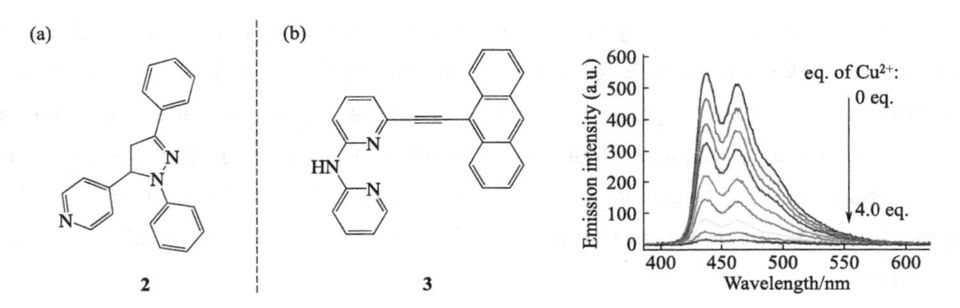

Fig.5.24 (a) Chemical structure of compound 2; (b) Left: chemical structure of probe 3; right: fluorescence emission spectral changes of probe 3 with Cu^{2+}

4. AND

AND is the logic type that many people recognize, perhaps because its human analogy embraces those universal values of cooperation and unity. Cultures old and new possess lines like "United we stand, divided we fall" and "(We are) better together". AND logic is contained in the word "synergy" which is popular in business circles. Any one "high" input by itself is powerless to produce the output, but the output comes alive when both "high" inputs are applied at the same time.

A sequence-activated AND logic dual-channel NIR fluorescent probe (Cy-S-CPT) is developed. The smart nanoprobe P(Cy-S-CPT) is composed of two functional components: an ionizable tertiary amine-containing diblock copolymer which renders an ultra-sensitive response to small pH differences between acidic tumour cells and blood, and a dual-channel NIR fluorescence component Cy-S-CPT for tracking the biothiol-triggered prodrug release in vivo (Fig.5.25).

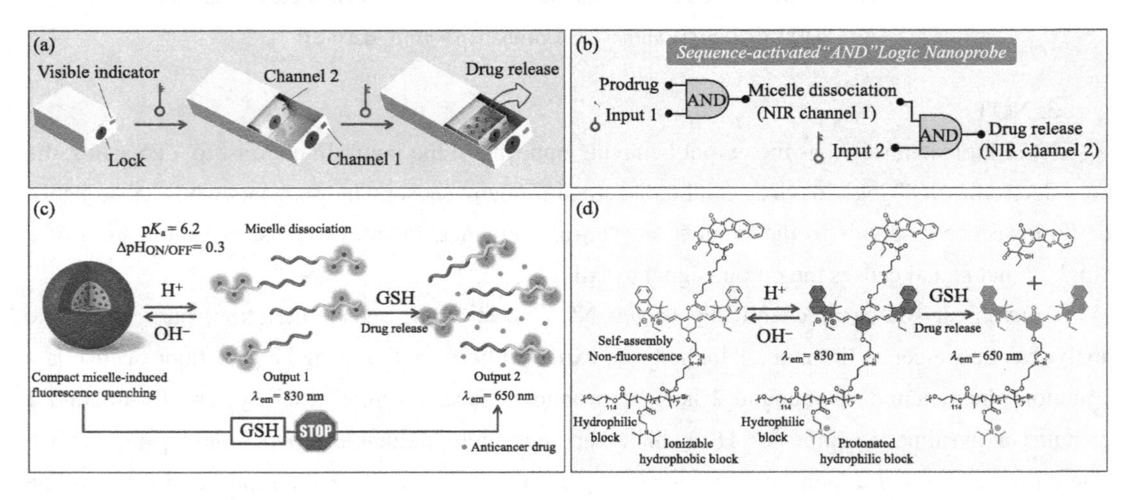

Fig.5.25 Sequence-activated AND logic dual-channel NIR fluorescent probe

(a and b) Logic circuit diagram of the sequence-activated AND logic nanoprobe for real-time tracking and programmable drug release via a dual NIR channel; (c and d) Schematic design of the sequence-activated AND logic dual-channel NIR theranostic nanoprodrug

An AND logic gate is built up so that the active output CPT is produced only if H^+ and GSH are both present as dual inputs. In particular, the programmable fluorescence response is carried out by the two sequence-dependent inputs. Specifically, output 1 is fluorescence at 830nm (NIR channel 1) which indicates whether the nanoprobe becomes activated and converts to a secondary form (disassembly). Subsequently coupled to the second input of GSH, the disassembled nanoprobe exhibits another readout fluorescence at 650nm (NIR channel 2) as output 2 and the concomitant release of active CPT [Fig.5.25(b) and (c)]. Taken all together, the sequential AND logic gate is successfully constructed, providing a promising strategy for selective sensing and targeted drug release.

5. OR

Previous sections on AND logic show our reliance on receptors binding their guests selectively so that each guest can serve as the correct input to each port of the molecular device. Now we head to the opposite extreme. Molecular OR logic might therefore seem an antithesis because most chemical research aims for selective reactions, except for notable exceptions, such as the non-selective reactions sought during photoaffinity labelling of enzymes and tagging of polymer beads for combinatorial chemistry. Indeed, well-behaved OR logic suggests perfectly unselective output production by different inputs. Unselective behaviour of a given receptor towards, say, two different guests allow us to approach OR logic devices containing a single receptor alone, i.e. double-input logic devices with a single port.

While old examples of molecular OR logic can be located, e.g. in **4**, the first deliberately designed gate of this type is **5** [Fig.5.26(a)]. "Fluorophore-spacer-receptor" PET system **5** has an aromatic amino acid receptor which acts as an electron donor towards the excited diarylpyrazoline fluorophore, resulting in negligible emission. This receptor is sufficiently unselective to Ca^{2+} or Mg^{2+} so that binding of either ion, supplied at a high enough concentration, blocks the electron-rich sites of the receptor and arrests PET. Each ion produces essentially identical extents of switching "on" [Fig.5.26(b)]. This similarity is due to essentially identical conformational changes produced upon complexation. Each ion-bound state effectively decouples the amine substituent from the oxybenzene unit so that PET is similarly suppressed. This also means that the charge density difference between the two cations is of secondary importance in these conformationally switchable systems. It is also notable that a single-receptor system is sufficient in this case to achieve a two-input logic gate.

Although semiconductors are essential for current computing devices, computation with molecules is no less realistic. Although the field is only 19 years old, it already has applications which are uniquely useful (e.g., tracking species and properties within cells and in tissue, improved medical diagnostics and screening for catalysts). We have to learn some elements of physiology, molecular biology, biochemistry, device physics, computer science, mathematics and aeronautical engineering (besides the several sub-disciplines of chemistry) in order to appreciate the gamut of molecular logic systems that we encounter.

(a)

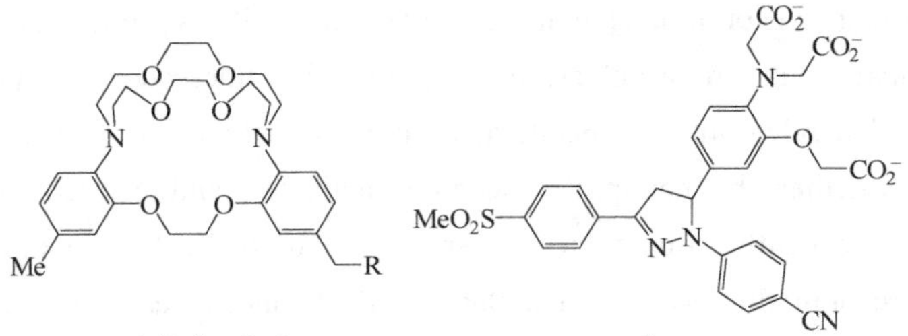

4: R=9-anthryl　　5

(b)

Truth table for **5**

Input 1 Ca^{2+}	Input2 Mg^{2+}	Output Fluorescence[b]
none	none	low (0.0042)
none	high (0.5mol·L^{-1})	high (0.24)
high (10^{-3} mol·L^{-1})	none	high (0.28)
high (10^{-3} mol·L^{-1})	high (0.5mol·L^{-1})	high (0.28)

[a]10^{-5} mol/L in water at pH 7.3.

[b]Quantum yields, λ_{exc} 389nm, λ_{em} 490nm.

Fig.5.26　(a) Chemical structure of 4 and 5; (b) Truth table for 5

Words and Expressions

molecular logic-based computation　分子逻辑计算

computational chemistry　计算化学

stimulus　*n.* 刺激；刺激物

absorbance　*n.* 吸光度；吸收率

emission intensity　发射强度

photoinduced electron transfer　光诱导电子转移

near-infrared region　近红外区

intramolecular charge transfer　分子内电荷转移

activate　*v.* 激活

disassembly　*n.* 解体；分解

identical　*adj.* 同一的；完全相同的

complexation　*n.* 络合；络合作用

Notes

➢ Molecular logic-based computation is not about computational chemistry, and computational chemistry is the subject that depends on computer programs based on quantum theory running on semiconductor-based hardware to provide information about atoms, molecules, materials and reactions.

参考译文：分子逻辑计算与计算化学无关，计算化学是一门学科，它依赖于基于量子理论的计算机程序，在基于半导体的硬件上运行，以提供关于原子、分子、材料和反应的信息。

➢ Optical responses (absorbance, emission intensity, emission anisotropy, ellipticity signals in circular dichroism spectra) serve as outputs since these are the most convenient for quick observation with inexpensive equipment or, sometimes, none at all.

参考译文：光学信号（吸光度、发光强度、发光各向异性、圆二色光谱中的椭圆率信号）可以作为输出信号，因为这些信号可以通过低成本的设备或不需要设备即可进行检测。

➢ Optical inputs, e.g., light dose, which cause photochemical reactions, are also useful. "High" and "low" signal levels can be taken as binary 1 and 0 respectively (or the opposite) whatever the inputs or outputs under discussion.

参考译文：能引发光化学反应的光学信号输入，比如光剂量，是非常有用的。无论是输入信号或输出信号，高和低的信号级别都可以设定为二进制的 1 和 0（或者反过来）。

Exercises

1. Discuss the following questions

(1) What is the difference between computational chemistry and molecular logic-based computation?

(2) What can serve as inputs and outputs in molecular logic-based computation?

(3) What are the inputs in AND logic probe Cy-S-CPT?

2. Translate the following into Chinese

(1) Molecular logic-based computation is about molecules and chemical systems which possess innate ability to compute, at least in a rudimentary way, like machines based on semiconductor transistors (or magnetic relays or mechanical abacuses in the past) or like people.

(2) Small molecules easily go to small vital spaces where semiconductor devices fear to tread.

(3) Any one "high" input by itself is powerless to produce the output, but the output comes alive when both "high" inputs are applied at the same time.

(4) In contrast to YES gates, the non-electronic NOT logic devices can be discerned in many sensors and reagents whose absorbance or fluorescence responds to the input in an "on-off" manner.

3. Translate the following into English

(1) 借用计算机科学家的语言来说，就是给分子器件一个输入信号，其相应地产生一个输出信号。

(2) 电子逻辑器件的显著特征就是输入信号的产生以及输出信号的观察都非常便利。

(3) 即使很多分子基材料无法产生单分子逻辑行为，但它们同样也是适用的。

(4) 质子可能是最简单的化学成分，但是它为许多概念的形成提供了一个非常有用的实验平台。

UNIT 5　Artificial Molecular Machines

1. Introduction

Machine, a glorious work of human civilization, is an efficient tool to liberate productivity, which is closely related to production and life. Diverse machines have greatly changed the way of human life.

As early as hundreds of millions of years ago, organisms have evolved a variety of biological machines in the body: DNA and RNA recording genetic information, ribosomes for protein assembly, actin that carries goods within the cell, molecular pumps that convert energy, and so on. These biological machines perform complex tasks all the time. Cells are like miniature factories, carrying biological machines and life activities. It can be said that the cell itself is a collection of molecular machines, and life depends on molecular machines.

Seeing the exquisite biological machine, people started to explore and imitate driven by curiosity, thus the concept of artificial molecular machine was put forward. Then a lot of work was carried out during last 6 decades, and the most representative event in the field is the 2016 Nobel Prize in Chemistry.

2. The Conception of Molecular Machine

Can humans use synthetic methods to create molecular-level machines that mimic the inherent biological machines? Can human beings manipulate these nano machines in the micro world as in the macro world? Can these nanomolecular machines replace biomolecular machines to perform complex biological behaviours in organisms? With these ideas and questions, R. Feynman gave a famous speech, "there is a lot of space at the bottom", proposing the possibility of building micro-machines on the molecular scale in 1959. However, during the period of Feynman, chemists did not have enough synthetic and analytical techniques to accurately create such molecules. Thanks to the development of synthetic chemistry and supramolecular chemistry, effective synthesis strategies for artificial molecular machines were provided, which resulted in rapid development in the field of molecular machines.

Since then, more and more chemists have paid attention to construct precisely regulated mechanical molecules, and developed into molecular machines that can do work at molecular scale. The 2016 Nobel Prize in Chemistry has been awarded to J. -P. Sauvage, J. F. Stoddart and B. L. Feringa for their outstanding work in the field of molecular machines (Fig.5.27).

3. Constructing Mechanical Bonds

Just as macro machines require various ways to combine parts or units, molecular machines also require ways of assembling parts together, and people first thought of mechanical bonds. What are mechanical bonds? According to Stoddart, a mechanical bond refers to a spatial entanglement between two or more molecular units, which makes it impossible for these molecular units to be separated by ways other than breaking chemical bonds. Molecules connected by mechanical bonds are called mechanically interlocked molecules. According to the molecular structure, these molecules can be divided into two basic types, catenane and rotaxane (Fig.5.28).

Ill: N. Elmehed. ©Nobel Media 2016
Jean-Pierre Sauvage
Prize share: 1/3

Ill: N. Elmehed. ©Nobel Media 2016
Sir J. Fraser Stoddart
Prize share: 1/3

Ill: N. Elmehed. ©Nobel Media 2016
Bernard L. Feringa
Prize share: 1/3

Fig.5.27 Portraits of researchers who won the 2016 Nobel Prize in Chemistry

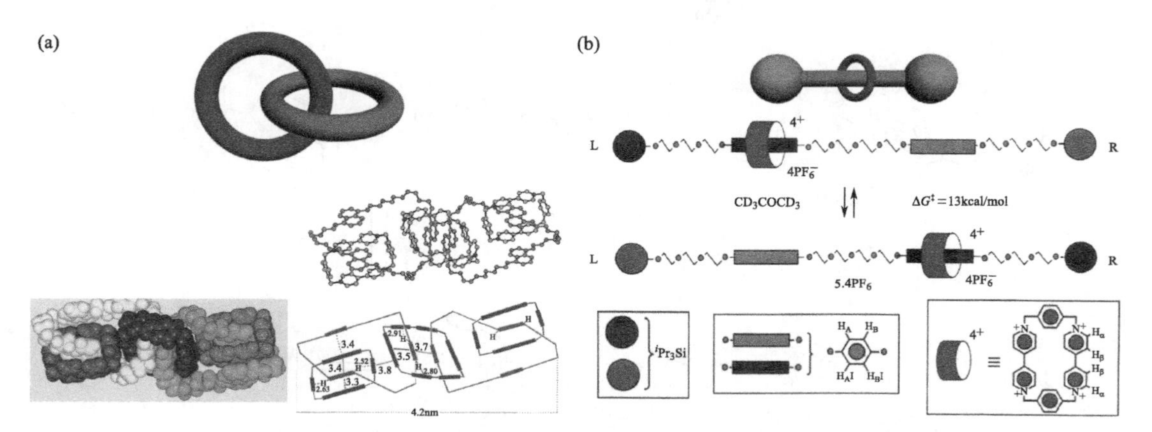

Fig.5.28 Structural diagrams and representative molecules of (a) catenane and (b) rotaxane

Catenane is a kind of interlocking structure of a buckle ring, in which the rotation of each ring will cause the relative position of the rings to change. Further, the number of rings and buckles can be expanded through topology to more complex interlocking structures.

The simplest rotaxane (i.e., a wheel containing two components) is composed of a dumbbell-shaped component and a set of cyclic components, where the ring can rotate or slide opposite the rod. Similarly, rotaxane can be developed into a complex system with a variety of interpenetrating, self-piercing and multi-dimensional systems through some specific template synthesis strategies.

4. Drive Mechanically Interlocked Molecules

After the introduction of mechanical bonds into molecules, these mechanically interlocked molecules tend to behave irregularly (i.e., in Brownian motion). They own a mechanical skeleton structure without the ability to control mechanical motion. Therefore, a deeper understanding and precise control of the interaction between the different components is needed to activate the molecular machine.

Sauvage was the first to find that the relative rotational motion of the two rings can be controlled by controlling the valence state of copper ions after the introduction of two different metal ion binding sites in the catenane. Such a discovery was the first prototype of non-biomolecular machine and was one of the most important reasons why Sauvage later won the Nobel Prize.

The first rotaxane-based molecular shuttle was reported by Stoddart in 1991. After a deep understanding of the non-covalent force between electron-deficient macrocyclic molecules and electron-rich guest molecules, Stoddart and his co-workers introduced two equivalent hydroquinone units and a cyclic component into the rod-shaped components to form a molecular shuttle. As recognition sites with a certain distance but equivalent, the two hydroquinone units make macrocyclic molecules bind to them with equivalent non-covalent forces, so the cyclic component could shuttle between the two recognition sites in a sliding way (Fig.5.29).

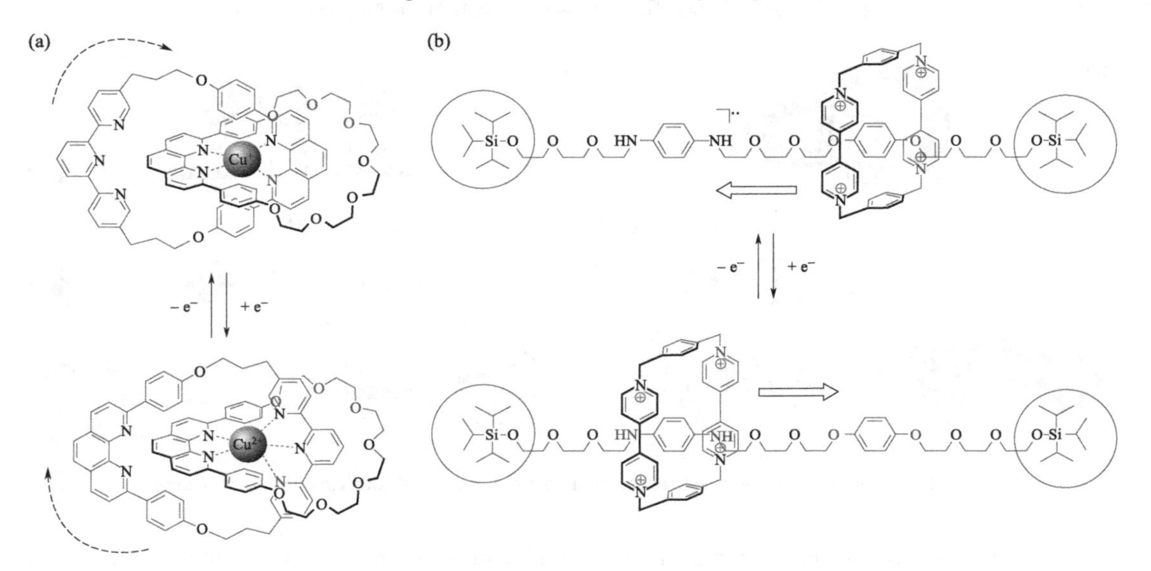

Fig.5.29 (a) Catenane and (b) rotaxane driven by redox

The discovery of the rotational motion in mechanically interlocked molecules and molecular shuttle marks the beginning of a milestone in the evolution of mechanically interlocked molecules to molecular switches and molecular machines.

5. From Molecular Switches to Molecular Machines

Once mechanically interlocked molecules could be controlled by controlling thermodynamic equilibria, chemists began to focus on how to make these molecular machine prototypes perform specific functions.

Although it is possible to make these mechanically bonded molecules "move", they are still not strictly molecular machines. Because the conversion of their structure is a transition between two thermodynamic equilibrium states, and the driving force is a change in chemical equilibrium (for example, redox reaction is the way of driving). Thus, bistable catenanes and rotaxanes may act as molecular switches, but they are not yet machines. The main difference between a molecular machine and a molecular switch is that the former can do work, while the latter is just a transition between two thermodynamic equilibria but cannot output work.

The representative functional molecular machines include molecular muscle, molecular elevator, molecular pump, molecular synthesis line and so on (Fig.5.30). In 2000, Sauvage constructed molecular muscles based on metal ion coordination, also known as molecular daisy chains. It has two stretching states just like muscle units, and there is a significant change in the molecular length corresponding to these two stretching states. If a certain weight is loaded at the ends, the molecular unit will be able to generate mechanical force and displacement under the condition of external stimulation (input energy), that is, the molecular could employ external chemical energy to do mechanical work. Similarly, for molecular elevators reported in 2004, if a "human" is placed on a platform composed of crown ether, then the elevator would raise the "him" by 0.7nm under acid-base stimulation, which is also a typical molecular machine that can do mechanical work.

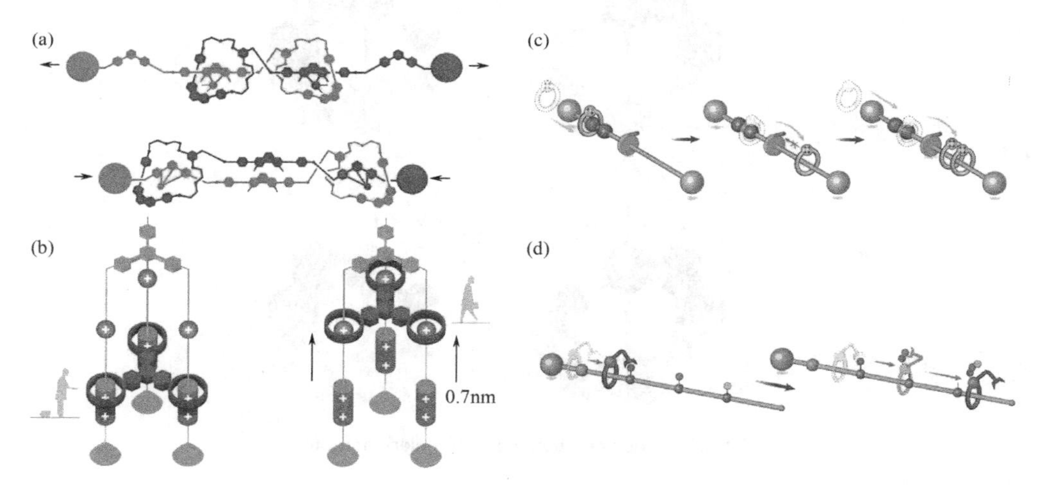

Fig.5.30 A schematic diagram of a representative functional molecular machine

(a) molecular muscle; (b) molecular elevator; (c) molecular pump; (d) molecular synthesis line

Molecular machines are not limited to doing mechanical work. The molecular machine reported by Stoddart can overcome the potential energy difference and transport small molecules from low to high concentrations just like the active transport of cell membranes, which is called molecular pumps. Molecular pumps can convert chemical energy into the potential energy of macrocyclic molecules by redox stimulation. Therefore, one of the major topics in the field of molecular machines is how to simulate the behaviour of biomacromolecules using artificial molecular machines. Another representative work is the artificial peptide synthesis machine. It can automatically synthesize peptide molecules under certain chemical stimulation.

6. Further, to Molecular Motor

Unlike machines based on supramolecular chemistry, molecular motor is based on sterically hindered olefins. The molecular motors originate from light-responsive olefin molecular switches that have been known to activate the *cis-trans* isomerization excited by ultraviolet light. The difference is that Feringa introduces several stereochemical centers into this classic molecular switch to control the stability in different states. A molecule with motor function is rationally designed: the group at one end of the double bond can undergo two *cis-trans* isomerism and two thermal relaxation processes under the excitation of ultraviolet light, thus performing a unidirectional rotation (Fig.5.31). It can convert light energy into mechanical energy, which is another milestone in the field of molecular machines. In the subsequent research, the molecular motor can drive the rotation of liquid crystal molecules, and even drive micron glass rods which are thousands of times larger than the motor itself. Molecular motors achieve the goal of employing microscopic molecular to do mechanical work on macroscopic objects.

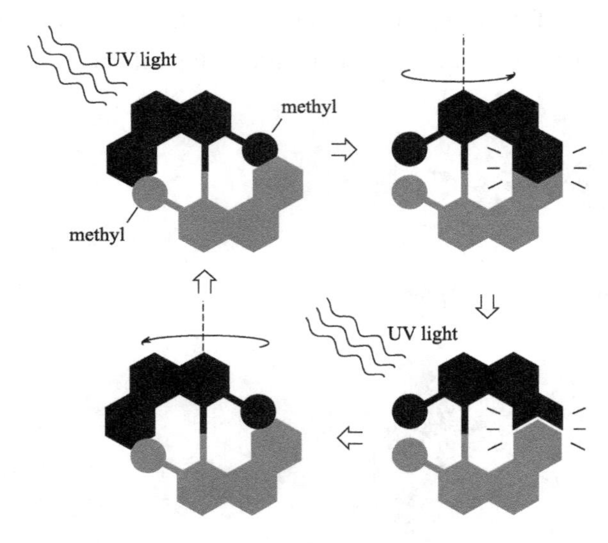

Fig.5.31 The operation mode of molecular motor

7. Epilogue

Feringa once said: maybe the power of chemistry is not only understanding, but also creation, creating molecules and substances that have never existed. Molecular machines are pioneering work for people to deeply understand and precisely control molecules. Applying supramolecular chemistry, especially supramolecular self-assembly process, to amplify the single molecular motion machines to macroscopic scale will be an important direction for the development of molecular machines, which will greatly promote the process of device and material of molecular machines.

Over the past 30 years, people have made remarkable achievements in the field of molecular machines, which have pushed human creativity in using chemical technology to a higher level. However, it is undeniable that there is still a long way to go to really apply this technology to practical applications. It also indicates that the basic research of molecular machines will continue in the future for quite a long time. In addition, the Nobel Prize in Chemistry will greatly promote the development of molecular machines, attract numerous scientists joining the field. As a feat that humans imitate and surpass nature, molecular machine is bound to have good prospects in materials, biology, medicine and other fields.

Words and Expressions

mimic *v.* 模仿
molecular machine 分子机器
artificial molecular machine 人工分子机器
exquisite *adj.* 精致的
inherent *adj.* 固有的；内在的；遗传的
micro *adj.* 极小的；基本的；微小的；微观的
macro *adj.* 大规模的；宏观的
organism *n.* 有机体；生物体
mechanical bond 机械键
entanglement *n.* 纠缠
mechanically interlocked molecule 机械互锁分子
catenane *n.* 索烃
rotaxane *n.* 轮烷
prototype *n.* 原型
equivalent *adj.* 相等的；等价的
molecular switch 分子开关
thermodynamic equilibria 热力学平衡
molecular motor 分子马达
sterically hindered olefin 位阻烯烃
microscopic *adj.* 微观的
macroscopic *adj.* 宏观的

Notes

➢ Thanks to the development of synthetic chemistry and supramolecular chemistry, effective synthesis strategies for artificial molecular machines were provided, which resulted in rapid development in the field of molecular machines.

参考译文：合成化学和超分子化学的发展，为人工分子机器提供了有效的合成策略，这也引发了分子机器领域的迅速发展。

➢ A mechanical bond refers to a spatial entanglement between two or more molecular units, which makes it impossible for these molecular units to be separated by ways other than breaking chemical bonds.

参考译文：机械键是指两个或多个分子单元之间的空间纠缠作用，这种作用使得这些分子单元不能以化学键断裂之外的方式分离。

➢ As recognition sites with a certain distance but equivalent, the two hydroquinone units make macrocyclic molecules bind to them with equivalent non-covalent forces, so the cyclic component could shuttle between the two recognition sites in a sliding way.

参考译文：作为有一定距离但等价的识别位点，两个对苯二酚单元使大环分子之间以等价的非共价力结合，因此环状组分能够以滑动的方式穿梭于两个识别位点之间。

➢ The main difference between a molecular machine and a molecular switch is that the former can do work, while the latter is just a transition between two thermodynamic equilibria but cannot output work.

参考译文：分子机器和分子开关之间的主要区别在于前者可以做功，而后者只能在两个热力学平衡之间转换而不能输出功。

Exercises

1. Discuss the following questions

(1) What are mechanical bonds?

(2) What is the difference between molecular machines and molecular switches?

(3) How is energy transformed in Feringa's molecular motors?

2. Translate the following into Chinese

(1) As early as hundreds of millions of years ago, organisms have evolved a variety of biological machines in the body: DNA and RNA recording genetic information, ribosomes for protein assembly, actin that carries goods within the cell, molecular pumps that convert energy, and so on.

(2) Human beings could manipulate these nano machines in the micro world as in the macro world.

(3) Chemists have paid attention to construct precisely regulated mechanical molecules, and developed into molecular machines that can do work at molecular scale.

(4) Catenane is a kind of interlocking structure of a buckle ring, in which the rotation of each ring will cause the relative position of the rings to change.

(5) These mechanically interlocked molecules tend to behave irregularly (i.e., in Brownian motion).

(6) Once mechanically interlocked molecules could be controlled by controlling thermodynamic equilibria, chemists began to focus on how to make these molecular machine prototypes perform specific functions.

(7) Molecular machines are pioneering work for people to deeply understand and precisely control molecules.

3. Translate the following into English

(1) 从那时起，越来越多的化学家开始关注构建精确调控的机械分子。

(2) 正如宏观机器需要各种方式来组合部件或单元一样，分子机器也需要将部件组装在一起，而人们首先想到的方式是机械键。

(3) 具有代表性的功能性分子机器包括分子肌肉、分子升降机、分子泵、分子合成线等。

(4) 因此，如何利用人工分子机器模拟生物大分子的行为是分子机器领域的主要课题之一。

(5) 分子马达实现了利用微观分子对宏观物体进行机械工作的目标。

UNIT 6 Carbon Neutral

Carbon neutral is the new gold. Nowadays, more and more countries pledge to become carbon neutral, net-zero or even climate positive. Terms like "carbon neutrality" "net-zero" or "climate positive" have been around for a while, but for the last couple of years, small startups to global corporations have integrated them, mainly for mainstream marketing purposes.

1. What is Carbon Neutral?

Carbon neutral was the *New Oxford American Dictionary's* word of the year in 2006 and since then, has been catapulted into the mainstream world. By definition, carbon neutral (or carbon neutrality) is the balance between emitting carbon and absorbing carbon emissions from carbon sinks. Or simply, eliminate all carbon emissions altogether. Carbon sinks are any systems that absorb more carbon than they emit, such as forests, soils and oceans.

According to the European Union Commission, natural sinks remove between 9.5Gt and 11Gt of CO_2 per year. To date, no artificial carbon sinks can remove carbon from the atmosphere on the necessary scale to fight global warming. Hence, to become carbon neutral, countries have two options: reducing drastically their carbon emissions to net-zero or balancing their emissions through offsetting and the purchase of carbon credits.

2. What Does It Mean to Become Carbon Neutral?

音频

Becoming carbon neutral is the new mantra of Wall Street and worldwide companies, but how to make it happen? It is impossible to generate zero-carbon emissions. Therefore, offsetting is a viable approach to become carbon neutral. Offsetting your carbon emissions sends a strong message to your community, that you are committed to paving the way for a sustainable future. The funds from neutralizing your carbon footprint will be providing low-carbon technology to communities most at risk of the impacts of climate change. However, you have to ensure that the offsetting project is transparent and involves local communities in the process.

3. What is the Difference between Carbon Neutral and Net-zero?

As established previously, carbon neutral and net-zero are two similar terms. In both cases, companies are working to reduce and balance their carbon footprint. When carbon neutral refers to balancing out the total amount of carbon emissions, net-zero carbon means no carbon is emitted from the get-go, so no carbon needs to be captured or offset. For example, a company's building running entirely on solar, and using zero fossil fuels can label its energy as "zero carbon".

However, when referring to "net-zero", it is crucial to specify net-zero carbon or emissions. On the contrary, net-zero emissions refer to the overall balance of greenhouse gas emissions (GHG) produced and GHG emissions taken out of the atmosphere. Even if the scientific concept is often applied to countries like US, China, it can also be used for organizations. In other words, net-zero describes the point in time where humans stop adding to the burden of climate-heating gases in the atmosphere.

4. Carbon Negative or Climate Positive: Doing More for the Planet

Carbon negative and climate positive are two similar terms. It occurs when a country removes or captures more CO_2 from the atmosphere than it even emits. Then, the country has a negative amount of carbon emissions and positively impacts the climate. Since carbon neutral or climate positive for companies becomes the new trend or new gold, some firms are already seeing further and tend to erase the totality of their historical footprint.

China, the world's largest emitter of carbon dioxide, has promised to become carbon neutral before 2060, and to begin cutting its emissions within the next ten years. President Xi Jinping made the ambitious pledge to a virtual audience of world leaders at a meeting of the United Nations General Assembly. The news came as a surprise to many researchers, even in China, who were not expecting such a bold target. It is the country's first long-term climate goal, and will require China to rein in CO_2 and probably other greenhouse-gas emissions to net-zero, which means offsetting gases that are released, for example by planting trees or capturing carbon and storing it underground.

In the wake of the announcement, several proposals come from influential research groups that work closely with the government on how China could reach neutrality before 2060. The plans differ in their details, but agree that China must firstly begin to generate most of its electricity from zero-emission sources, and then expand the use of this clean power wherever possible, for example switching from petrol-fuelled cars to electric ones. It will also need technologies that can capture CO_2 released from burning fossil fuels or biomass and store it underground, known as carbon capture and storage (CCS).

(1) Renewable Boost

For China to achieve its target, electricity production would need to more than double, to 15034 terawatt hours by 2060, largely from clean sources. This growth would be driven by a massive ramp up of renewable electricity generation over the next 40 years, including a 16-fold increase in solar and a 9-fold increase in wind (Fig.5.32). To replace coal-fired power generation, nuclear power would need to increase 6-fold, and hydroelectricity to double.

Fig.5.32 China massively increase its solar and wind capacity to become carbon neutral by 2060

Fossil fuels, including coal, oil and gas, would still account for 16% of energy consumed, so would need to be paired with CCS or offset by new forest growth and technologies that can suck CO_2 directly out of the atmosphere.

Under their plan, emissions would continue to rise, from 9.8 gigatonnes of CO_2 in 2020 to around 10.3 gigatonnes in 2025. They will then plateau for five to ten years before dropping steeply after 2035, to reach net-zero by 2060.

But shifting China's economy away from its dependence on fossil fuels in such a short time will be expensive, says Levine. Coal-fired power accounts for almost 65% of the country's electricity generation, with more than 200 new coal-fired power stations planned or under construction. "There will be tremendous opposition" from industries that rely on fossil fuels, he says.

A major cost will be the energy storage required to integrate wind and solar at such a scale. But battery storage has become cheaper over the past decade, and that could bring costs down. If trends in the cost of renewable technology continue, more than 60% of China's electricity could come from non-fossil fuels by 2030. That is quite encouraging. However, ensuring stable operation of the electricity grid, given the intermittent nature of wind and solar power, will be another challenge.

(2) Ramp Up Nuclear

Other teams have also envisioned a carbon neutral future for China. Beijing would see emissions peak as soon as 2022, at around 10 gigatonnes of CO_2, followed by a steep drop to net-zero by 2050. To achieve this, electricity production would double to 14800 terawatt hours by 2050. This output is similar to that in Zhang's model, but would be generated largely by nuclear power (28%), followed by wind (21%), solar (17%), hydropower (14%) and biomass (8%). Coal and gas would make up 12% of electricity production.

This means that China's nuclear capacity — currently 49 gigawatts across some 50 nuclear power plants — would need to increase 5-fold, to 554 gigawatts by 2050, through rapid construction of new sites. Nuclear power would supply a more consistent base load of power than solar and wind.

(3) Cut Down Coal

Jiang's analysis also points to another crucial sticking point among researchers — the role of CCS. The model proposes that some 850 gigawatts of power generated from coal, gas and biofuels could be fitted with technologies that capture and store carbon emissions.

A stringent climate target requires substantial deployment of CCS. But this would require significant investment, because China currently has only one large CCS facility in operation, at an oil field. Seven more facilities are being planned or built. CCS would allow China to continue using some coal-fired power in the long term, but some researchers say the technology is still very expensive, which limits its application. Many researchers think China should stop building new coal-fired power plants. Existing plants will reach the end of their life before the neutrality deadline in 40 years.

(4) Decisions Ahead

The path the country will follow to reach neutral emissions will probably become clearer in the coming months. China, like all nations that have signed the 2015 Paris climate agreement, is obliged to submit increased emissions-reduction targets before the end of the year.

At the Beijing meeting on 12 October 2020, researchers proposed increasing the target for the proportion of China's energy that is produced by non-fossil fuels in 2030 from 20% to 25%, and bringing the 20% goal forward to 2025. The neutrality target includes all greenhouse gases, including methane. Officials are in the process of drafting the country's latest five-year plan for social and economic development, which will be released in March and is expected to include policies to achieve neutrality. A detailed breakdown of energy and climate targets will show how serious China is about reaching its carbon neutrality goal.

Words and Expressions

carbon neutral 碳中和
net-zero carbon emission 净零碳排放
climate positive 气候正效应
carbon sink 碳存储
greenhouse gas emission 温室气体排放
petrol-fuelled car 石油燃料汽车
carbon capture and storage (CCS) 二氧化碳捕获和储存
renewable electricity generation 可再生能源发电
fossil fuel 化石燃料
ramp up 强化
neutral emission 中和排放
non-fossil fuel 非化石燃料

Notes

➢ Terms like "carbon neutrality" "net-zero" or "climate positive" have been around for a while, but for the last couple of years, small startups to global corporations have integrated them, mainly for mainstream marketing purposes.

参考译文："碳中和""净零"或"气候正效应"等术语已经存在一段时间了，但在过去几年中，小型初创企业和跨国公司已经将它们整合起来，主要用于主流营销。

➢ By definition, carbon neutral (or carbon neutrality) is the balance between emitting carbon and absorbing carbon emissions from carbon sinks. Or simply, elimination all carbon emissions altogether. Carbon sinks are any systems that absorb more carbon than they emit, such as forests, soils and oceans.

➢ 参考译文：根据定义，碳中和是碳排放和由碳汇产生的碳吸收之间的平衡。或者简单地说，完全消除所有的碳排放。碳汇是任何吸收碳多于排放碳的系统，例如森林、土壤和海洋。

➢ To date, no artificial carbon sinks can remove carbon from the atmosphere on the necessary scale to fight global warming. Hence, to become carbon neutral, countries have two options:

reducing drastically their carbon emissions to net-zero or balancing their emissions through offsetting and the purchase of carbon credits.

➢ 参考译文：迄今为止，任何人工的碳存储都无法规模化地从大气中去除碳以应对全球变暖的情况。因此，要实现碳中和，有两种选择：将碳排放大幅减少至净零，或通过抵消和购买碳信用来平衡排放。

➢ In other words, net-zero describes the point in time where humans stop adding to the burden of climate-heating gases in the atmosphere.

参考译文：换而言之，净零碳排放为停止人类增加大气中的温室气体提供了时间点。

Exercises

Discuss the following questions

(1) How China could be carbon neutral by mid-century?

(2) What is the difference between carbon neutral, net-zero and climate positive?

References

1. Bettelheim F A, Brown W H, Campbell M K, et al. Introduction to general, organic, and biochemistry [M]. 12th ed. Chicago: Cengage Learning, 2019.

2. Lind G. Teaching inorganic nomenclature: a systematic approach[J]. Journal of Chemical Education, 1992, 69(8): 613-614.

3. Ma Y X. Special English for chemistry[M]. Tianjin: Tianjin University Press, 2005.

4. Crone R A. A History of color: The evolution of theories of lights and color[M]. Netherlands: Springer, 1999: 2-10.

5. Flesch P G. Light and light sources: High-intensity discharge lamps[M]. Berlin: Springer, 2006: 17-50.

6. Ozaki Y, Huck C, Tsuchikawa S, et al. Near-infrared spectroscopy theory, spectral analysis, instrumentation, and applications: Theory, spectral analysis, instrumentation, and applications[M]. Singapore: Springer, 2021: 3-5.

7. Theophanides T T. Infrared spectroscopy- anharmonicity of biomolecules, crosslinking of biopolymers, food quality and medical applications[M]. London: InTech, 2015: 5-9.

8. Higson S. Analytical chemistry[M]. Oxford: Oxford University Press, 2013: 55-63.

9. Fleming L, Williams D H. Spectroscopic methods in organic chemistry[M]. New York: McGraw Hill, 2019: 1-7.

10. Chapeaurouge A, Bigler L, Schäfer A, et al. Correlation of stereoselectivity and ion response in electrospray mass-spectrometry. Electrospray ionization-mass spectrometry as a tool to predict chemical behaviour[J]. Journal of the American Society for Mass Spectrometry, 1995, 6(3): 207-214.

11. Casey M, Leonard J, Lygo B, et al. Advanced practical organic chemistry[M]. Boston: Springer, 1990: 141-187.

12. Sharp J T, Gosney I, Rowley A. Practical organic chemistry: A student handbook of techniques[M]. Dordrecht: Springer, 1989: 54-113.

13. Hamilton A E, Buxton A M, Peeples C J, et al. An operationally simple aqueous Suzuki-Miyaura cross-coupling reaction for an undergraduate organic chemistry laboratory[J]. Journal of Chemical Education, 2013, 90(11): 1509-1513.

14. Wan H, Djokic N, Brown B A, et al. Using whiskey-flavoring compounds to teach distillation and IR spectroscopy to first-semester organic chemistry students[J]. Journal of Chemical Education, 2014, 91(1): 123-125.

15. Taber D F, Hoerrner R S. Column chromatography: Isolation of caffeine[J]. Journal of Chemical Education, 1991, 68(1): 73.

16. Yang Q, Canturk B, Gray K C, et al. Evaluation of potential safety hazards associated with the Suzuki-Miyaura cross-coupling of aryl bromides with vinylboron species[J]. Organic Process Research and Development, 2018, 22: 315.

17. Parmentier M, Wagner M, Wickendick R, et al. A general kilogram scale protocol for Suzuki-Miyaura cross-coupling in water with TPGS-750M surfactant[J]. Organic Process Research and Development, 2020, 24(8): 1536-1542.

18. Pollak P. Fine chemicals: The industry and the business[M]. 2nd ed. New York: John Wiley, 2007: 3-7.

19. Cybulski A, Sharma M M, Sheldon R A, et al. Fine chemicals manufacture: Technology and engineering [M]. Amsterdam: Elsevier, 2001.

20. Geng Y, Ling S, Huang J, et al. Multiphase microfluidics: Fundamentals, fabrication, and functions[J]. Small, 2020, 16(6): e1906357.

21. Kim H, Min K I, Inoue K, et al. Submillisecond organic synthesis: Outpacing Fries rearrangement through microfluidic rapid mixing[J]. Science, 2016, 47(6286): 691-694.

22. Hunger K. Industrial dyes: Chemistry, properties, applications[M]. New York: Wiley, 2003.

23. Clark M. Handbook of textile and industrial dyeing: Principles, processes and types of dyes[M]. Cambridge: Woodhead Publishing, 2011.

24. Holmberg K, Jonsson, B, Dronberg B, et al. Surfactants and polymers in aqueous solution[M]. 2nd ed. New York: Wiley, 2003.

25. Narula A. Fragrance and attraction[M]. Washington DC: ACS Publications, 2019.

26. Camille G. The practice of medicinal chemistry[M]. 2nd ed. Burlington: Elsevier, 2003.

27. Lin G, You Q, Cheng J. Chiral drugs: Chemistry and biological action[M]. New York: Wiley, 2011.

28. Dam G, Themelis G, Crane L, et al. Intraoperative tumor-specific fluorescence imaging in ovarian cancer by folate receptor-alpha targeting: First in-human results[J]. Nature Medicine, 2011, 17(10): 1315-1319.

29. Hu Z, Fang C, Li B, et al. First-in-human liver-tumour surgery guided by multispectral fluorescence imaging in the visible and near-infrared-I/II windows[J]. Nature Biomedical Engineering, 2020, 4(3): 1-13.

30. Guo Z, Shao A, Zhu W. Long wavelength AIEgen of quinoline-malononitrile[J]. Journal of Materials Chemistry C, 2016, 4: 2640.

31. Gu K, Qiu W, Guo Z, et al. An enzyme-activatable probe liberating AIEgens: On-site sensing and long-term tracking of β-galactosidase in ovarian cancer cells[J]. Chemical Science, 2019, 10: 398.

32. Fu W, Yan C, Guo Z, et al. Rational design of near-infrared aggregation-induced-emission-active probes: In situ mapping of amyloid-beta plaques with ultrasensitivity and high-fidelity[J]. Journal of the American Chemical Society, 2019, 141: 3171.

33. Nakatani K, Piard J, Yu P, et al. Photochromic materials[M]. New York: Wiley, 2016: 1-45.

34. Irie M, Fukaminato T, Matsuda K, et al. Photochromism of diarylethene molecules and crystals: Memories, switches, and actuators[J]. Chemical Reviews, 2014, 114(24): 12174-12277.

35. Irie M. Diarylethenes for memories and switches[J]. Chemical Reviews, 2000, 100(5): 1685-1716.

36. Zhu W, Yang Y, Rémi Métivier, et al. Unprecedented stability of a photochromic bisthienylethene based on benzobisthiadiazole as an ethene bridge[J]. Angewandte Chemie International Edition, 2011, 50 (46): 10986-10990.

37. Li W, Jiao C, Li X, et al. Separation of photoactive conformers based on hindered diarylethenes: Efficient modulation in photocyclization quantum yields[J]. Angewandte Chemie, 2014, 53 (18): 4603-4607.

38. Hagfeldt A, Boschloo G, Sun L, et al. Dye-sensitized solar cells[J]. Journal of Photochemistry &

Photobiology C Photochemistry Reviews, 2003, 4(2): 145-153.

39. Gratzel M. Recent advances in sensitized mesoscopic solar cells[J]. Accounts of Chemical Research, 2009, 42(11): 1788-1798.

40. Wu Y, Zhu W. Organic sensitizers from D-π-A to D-A-π-A: Effect of the internal electron-withdrawing units on molecular absorption, energy levels and photovoltaic performances[J]. Chemical Society Reviews, 2013, 42: 2039.

41. Wu Y, Zhu W H, Zakeeruddin S M, et al. Insight into D-A-π-A structured sensitizers: A promising route to highly efficient and stable dye-sensitized solar cells[J]. ACS Applied Materials & Interfaces, 2015, 7(18): 9307-9318.

42. Silva A. Molecular logic-based computation[M]. Cambridge: RSC Publishing, 2012.

43. Shi L, Yan C, Ma Y, et al. In vivo ratiometric tracking of endogenous β-galactosidase activity using an activatable near-infrared fluorescent probe[J]. Chemical Communications, 2019, 55(82): 12308-12311.

44. Xie Y, Ding Y, Li X, et al. Selective, sensitive and reversible "turn-on" fluorescent cyanide probes based on 2, 2′-dipyridylaminoanthracene-Cu^{2+} ensembles[J]. Chemical Communications, 2012, 48 (94), 11513-11515.

45. Yan C, Guo Z, Liu Y, et al. A sequence-activated and logic dual-channel fluorescent probe for tracking programmable drug release[J]. Chemical Science, 2018, 9 (29): 6176-6182.

46. Stoddart J F. Molecular machines[J]. Accounts of Chemical Research, 2001, 34 (6): 410-411.

47. Anelli P L, Spencer N, Stoddart J F. A molecular shuttle[J]. Journal of the American Chemical Society, 1991, 113 (13): 5131-5133.

48. Amabilino D B, Ashton P R, Reder A S, et al. Olympiadane[J]. Angewandte Chemie International Edition in English, 1994, 33(12): 1286-1290.

49. Livoreil A, Dietrich-Buchecker C O, Sauvage J P. Electrochemically triggered swinging of a [2]-catenate [J]. Journal of the American Chemical Society, 1994, 116 (20): 9399-9400.

50. Jimenez M C, Dietrich-Buchecker C, Sauvage J P. Towards synthetic molecular muscles: Contraction and stretching of a linear rotaxane dimer[J]. Angewandte Chemie International Edition, 2000, 39(18): 3284-3287.

51. Badjić J D, Balzani V, Credi A, et al. A molecular elevator[J]. Science, 2004, 303 (5665): 1845-1849.

52. Cheng C, Mcgonigal P R, Liu W G, et al. Energetically demanding transport in a supramolecular assembly[J]. Journal of the American Chemical Society, 2014, 136(42): 14702-14705.

53. Lewandowski B, Bo G D, Ward J W, et al. Sequence-specific peptide synthesis by an artificial small-molecule machine[J]. Science, 2013, 339(6116): 189-193.

54. Koumura N, Zijlstra R W J, van Delden R A, et al. Light-driven monodirectional molecular rotor[J]. Nature, 1999, 401(6749): 152-155.

55. Eelkema R, Pollard M M, Vicario J, et al. Nanomotor rotates microscale objects[J]. Nature, 2006, 440 (7081): 163.

56. Mallapaty S. How China could be carbon neutral by mid-century[J]. Nature, 2020, 586(7830): 482-483.